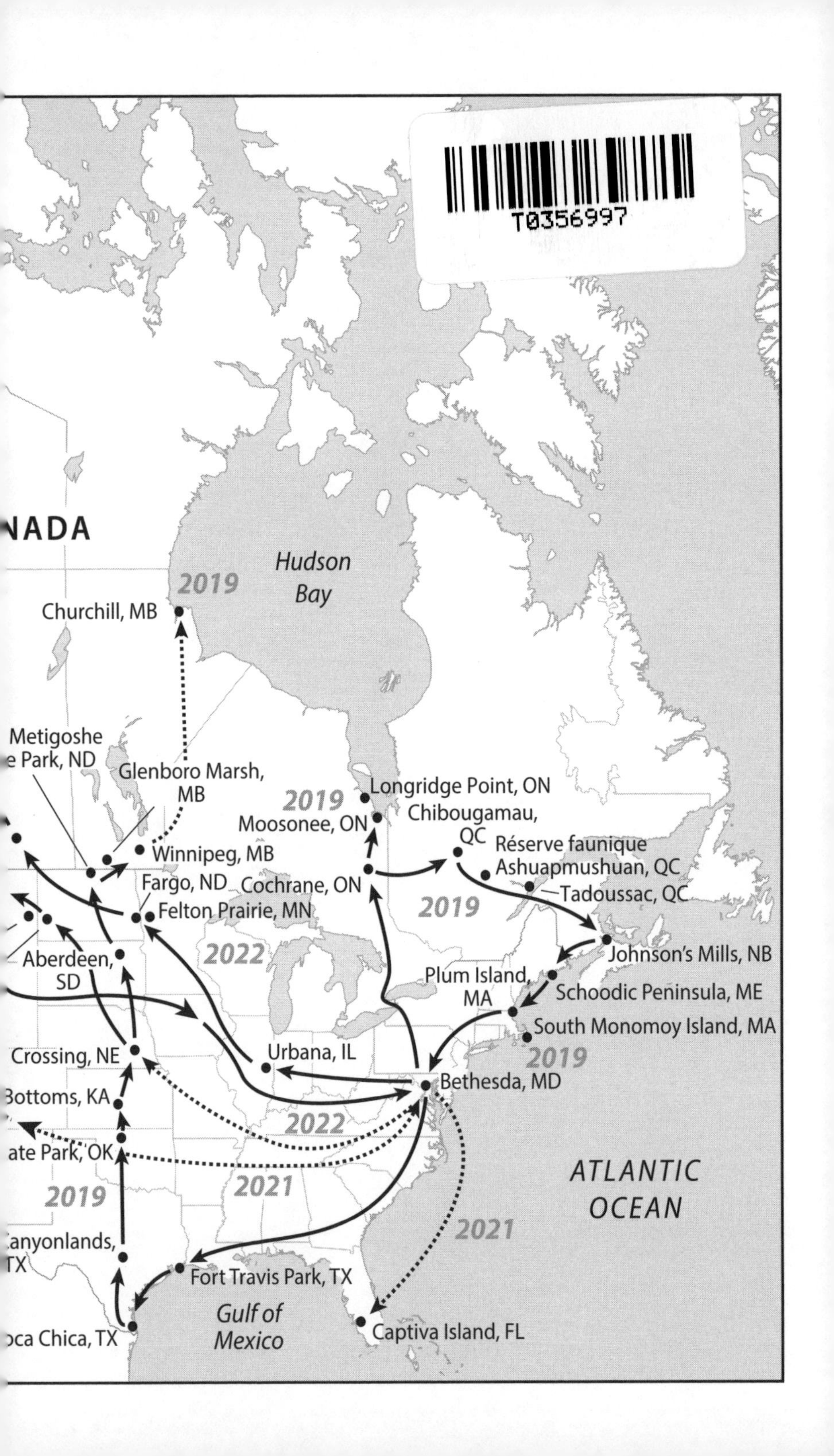
T0356997
NADA
Hudson Bay
2019
Churchill, MB
Metigoshe
e Park, ND
Glenboro Marsh, MB
Winnipeg, MB
Fargo, ND
Felton Prairie, MN
Aberdeen, SD
Crossing, NE
Bottoms, KA
ate Park, OK
2019
Canyonlands, TX
oca Chica, TX
Longridge Point, ON
2019
Moosonee, ON
Chibougamau, QC
Réserve faunique Ashuapmushuan, QC
Tadoussac, QC
Cochrane, ON
2019
2022
Johnson's Mills, NB
Plum Island, MA
Schoodic Peninsula, ME
South Monomoy Island, MA
2019
Urbana, IL
Bethesda, MD
2022
2021
ATLANTIC OCEAN
2021
Fort Travis Park, TX
Gulf of Mexico
Captiva Island, FL

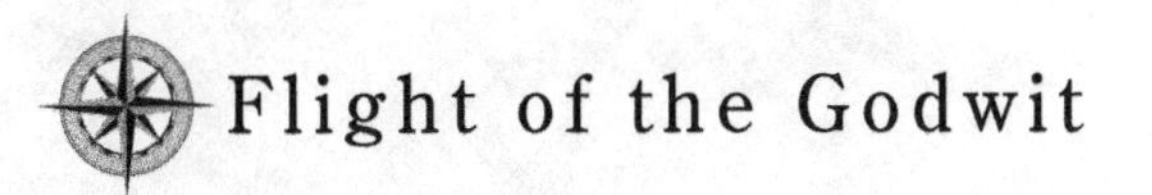
Flight of the Godwit

Flight of the Godwit

TRACKING EPIC SHOREBIRD MIGRATIONS

Bruce M. Beehler
Illustrations by Alan Messer

Smithsonian Books
Washington, DC

This book may be purchased for educational, business, or sales promotional use. For information, please write to the Special Markets Department at the address listed below.

Published by Smithsonian Books
PO Box 37012, MRC 513
Washington, DC 20013
smithsonianbooks.com
Director: Carolyn Gleason
Senior Editor: Jaime Schwender

Edited by Carrie Love
Designed by Broad Street Books
Endpapers and maps by Bill Nelson

Library of Congress Cataloging-in-Publication Data available upon request.
Hardcover ISBN: 978-1-58834-787-9
eBook ISBN: 978-1-58834-788-6
Printed in the United States of America
29 28 27 26 25   5 4 3 2 1

Title page: A flock of Hudsonian Godwits migrates over snow-clad Yukon mountains.

Dedication page: An Upland Sandpiper in flight.

Chapter opener image: A mixed flock of shorebirds on the shore, including a Hudsonian Godwit and a Whimbrel.

For Carol, Grace, Andrew, Cary, Chris, Mercy, and Zadie

And to the memory of artist and naturalist Alan Messer

How to Read This Book

In this book, I recount my journeys following the Hudsonian Godwit and a group of shorebirds I call "the Magnificent Seven." While my travels are the focus of the book, I have also included interesting information about the natural history of shorebirds, and I delve into the science related to these birds with technical sections peppered throughout the book, as well as a chapter on the wintering habits of shorebirds and one on their great migratory prowess. I know that not everyone will want to start with natural history: perhaps you already know all about shorebirds, and you want to dive right in with my travels—or perhaps you can't wait to find out the secrets of how these birds manage their long-distance migrations. With this in mind, the book is split into three parts:

Part I: We Go Way Back
Part II: Out on the Road
Part III: Until Next Year

Part I is all about history: my history with shorebirds and the natural history of the shorebirds—from their lineages and evolution to their behavior and appearance (and everything in between). If you want to start at the beginning, so to speak, start here. Part II is where we dive into my road trips in search of these incredible birds. If you are eager to get out on the road with me, flip to this part now! Part III begins with what we didn't learn about these birds earlier: what they do during the winter months. It ends with a description of the cutting-edge science that seeks to explain the great migratory powers of the shorebirds. If you are in a hurry to discover the secrets of migration, you might want to start at the end of the book and then jump back. However you decide to read the book, I hope that you are left with a better understanding of our shorebirds and some insight into why I so greatly admire these birds.

Contents

Featured Shorebirds

Below are the North American shorebird species featured in this book's narrative. They are listed following the systematic sequence of the North American Checklist of the American Ornithological Society.

- ☐ Upland Sandpiper*
- ☐ Bristle-thighed Curlew*
- ☐ Whimbrel*
- ☐ Eskimo Curlew†
- ☐ Long-billed Curlew*
- ☐ Bar-tailed Godwit*
- ☐ Hudsonian Godwit*
- ☐ Marbled Godwit*
- ☐ Ruddy Turnstone
- ☐ Red Knot
- ☐ Surfbird
- ☐ Stilt Sandpiper
- ☐ Sanderling
- ☐ Dunlin
- ☐ Purple Sandpiper
- ☐ Baird's Sandpiper
- ☐ Least Sandpiper
- ☐ White-rumped Sandpiper
- ☐ Buff-breasted Sandpiper
- ☐ Pectoral Sandpiper
- ☐ Semipalmated Sandpiper
- ☐ Western Sandpiper
- ☐ Short-billed Dowitcher
- ☐ Long-billed Dowitcher
- ☐ American Woodcock
- ☐ Wilson's Snipe
- ☐ Spotted Sandpiper
- ☐ Solitary Sandpiper
- ☐ Wandering Tattler
- ☐ Lesser Yellowlegs
- ☐ Greater Yellowlegs
- ☐ Willet
- ☐ Wilson's Phalarope
- ☐ Red-necked Phalarope
- ☐ Red Phalarope

* Members of the Magnificent Seven

† Extinct

Introduction

The Hudsonian Godwit is the one North American shorebird whose future still seems uncertain.... The flock of twenty-four I watched (with Robert Clem) one misty August afternoon of 1963 on the salt flats of Monomoy Island represented a large percentage [of the species] *that occur on the coasts of the Atlantic in the fall.*

—Peter Matthiessen, *The Wind Birds*

THE SEASONAL MOVEMENT OF BIRDS has fascinated philosophers, naturalists, and scientists for centuries. Much that was written in the early days of natural history was nonsense, based on long-repeated myths rather than direct observation in the field (no, swallows do not bury themselves in the mud of lakes for the winter as hypothesized by Olaus Magnus in the sixteenth century). In the nineteenth century, bird migration was mainly of interest because it offered the opportunity to shoot birds in large numbers for local consumption and for sale to city restaurants. Ornithologist and artist John James Audubon claimed to have attached silver threads to the legs of Eastern Phoebes to document their seasonal coming and going from his family home in Pennsylvania. In doing this, he became the first-known bird bander—a century ahead of the widespread practice of banding migrant birds to determine where they went each autumn.

In the 1940s, US scientists began to study bird migration in earnest. George Lowery, an ornithologist and professor of zoology at Louisiana State University, began studying spring migration by using a telescope to watch migrating birds passing across the face of the full moon at night. He was the first to postulate that large numbers of songbirds were flying north from the Yucatán directly across the Gulf of Mexico to the coast of Louisiana, an overwater distance of 600 miles, rather than following the "safer" overland route via the coast of Mexico. This was doubted by the ornithological establishment at the time, who could not believe small birds were able to make such a flight. Lowery's student, Sidney Gauthreaux, used weather radar in the 1960s to confirm Lowery's surmise.

Using several data sources, including the regular trapping of individuals that touched down on Bermuda, ornithologists I. C. T. Nisbet and colleagues determined that Blackpoll and Connecticut Warblers in autumn flew south over the Atlantic on multiday nonstop flights to wintering ranges in Amazonia. Biologists Stephen Emlen and Roswitha and Wolfgang Wiltschko used ingenious laboratory experiments deploying projected star patterns and mobile magnets to show that birds could use stellar movements as well as Earth's magnetism as navigational cues in migration. Major advances in technology and miniaturization of radio transmitters, nanotags, and geolocators have allowed scientists today to intimately follow the seasonal lives of individual avian migrants, with surprising and jaw-dropping results. Shorebirds have featured prominently in prime examples of epic long-distance migrations. This book will focus on these shorebirds and their remarkable travels.

In 2019, I followed up my fieldwork tracking the spring migration of wood warblers with a similar project on North American shorebirds, focusing on the larger and showier members of the family, especially the godwits and their close relatives. Researchers were only just beginning to discover the details of the prodigious travels of these wonderful birds. By visiting the breeding, staging (sites where birds rest during migration), and stopover locations of the Hudsonian Godwit, I got to travel across North America from the Mexican border north to Atlantic Canada, Hudson Bay, the Arctic Ocean, and the Bering Sea, encountering some of the wildest and most beautiful parts of our vast continent. I chose the Hudsonian Godwit as the flagship species for this project because it is a super-migrator (a species known to migrate long distances, often crossing substantial oceanic barriers in multiday overwater flights) whose migratory spring and autumn pathways led me across nearly every state and province of eastern, central, and northern North America. The godwits are arguably the most beautiful of our shorebirds, and among the very largest, as well as being the birds capable of the longest sustained migratory flights. This makes them among the most elusive for bird-watchers and therefore a more challenging subject for my project—a factor that made my journey even more interesting, permitting me here in this book to offer insights for fellow birders who may aspire to carry out similar adventurous travels (which I heartily recommend).

In spring, the northward-traveling Hudsonian Godwits depart southern South America, flying north over the eastern Pacific, crossing Central America, then passing north across the Gulf of Mexico, and making landfall in eastern Texas in April and May. From there they leapfrog northward, stopping as needed to rest and feed. Males move ahead of females to arrive on territory early to lock up a good patch of breeding habitat. In Manitoba, the migrating populations diverge. One group heads northwestward to Alaska and the Mackenzie River Delta in Arctic Canada, and the other continues northward to the western shore of Hudson Bay.

The migratory movements of the Hudsonian Godwit were the primary catalyst for my travels, but on this trip, I also followed the special assemblage of shorebirds I call "the Magnificent Seven," which includes the largest, most beautiful, and farthest-flying members of the family: Hudsonian Godwit, Bar-tailed Godwit, Marbled

Godwit, Whimbrel, Long-billed Curlew, Bristle-thighed Curlew, and Upland Sandpiper. These are North America's shorebird royalty. The Bar-tailed Godwit nests in Alaska and Siberia and winters in the Southwest Pacific. It is the greatest of overwater migrators. The adult male is handsomely plumaged, much like the Hudsonian Godwit. The Marbled Godwit is the giant among godwits, cinnamon-toned with abundant black flecking and a large pink-based, black-tipped, and upturned bill (the main feature of the godwits). The Long-billed Curlew, with its stupendous decurved bill, looks like an oversized Marbled Godwit, with buff and cinnamon plumage. It is our largest shorebird. The Whimbrel, a handsome and medium-sized curlew, is the most common and widespread of this group. It nests in the Arctic and sub-Arctic Canada and Alaska and winters from the southern United States to southern South America (the Eurasian population is now considered a distinct species by some authorities). The Bristle-thighed Curlew, nearly identical to the Whimbrel, is our rarest shorebird, with a tiny breeding range in western Alaska's uplands. It winters in the South Pacific. The Upland Sandpiper, looking like a short-billed curlew, has a similar dark-flecked plumage and a lovely and melancholy flight song. It summers across North America's prairies and winters in South America. I deem these seven species royalty for their beautiful plumages, large size, entrancing habits, and migratory prowess. Their swagger reminds me of the protagonists in Akira Kurosawa's 1954 film *Seven Samurai* (remade in the United States as *The Magnificent Seven*).

This new field trip replicated the travel model I deployed for the warbler migration studies I recounted in *North on the Wing*. My travels were confined to the United States and Canada, via solo travel by car and tent camping in parks and reserves. Contemporary eBird reports of godwits and other members of the Magnificent Seven informed my daily movements (eBird.org is a database of bird observations created by the Cornell Lab of Ornithology, and its online app enables birders to contribute georeferenced and dated

An adult Upland Sandpiper on its breeding habitat.

bird observations as well as examine these records in a convenient, map-based format—this system is invaluable for birders wishing to track down particular species). Even in countryside unknown to me, eBird reports helped orient my travels to where the birds were on a day-to-day basis. Thank you very much, Cornell Lab of Ornithology!

Between spring 2019 and spring 2023, I made 13 road trips in search of shorebirds. I logged 223 road days, traveling by car 35,087 miles, by airplane 14,614 miles (plus two helicopter charters), and by train 378 miles. I spent 108 days tent camping while traversing 37 states and 9 Canadian provinces and territories. I visited the breeding grounds of all three North American godwits, all the Magnificent Seven, and the important autumnal staging site of the Hudsonian Godwit on the western shore of James Bay in Canada. During this time I dealt with the vagaries of the COVID-19 pandemic, circumventing emergency procedures and dodging state and federal legal restrictions as necessary to be out with the birds. My objective was to spend as much time as possible in the field with the Hudsonian Godwit, the two other North American godwit species, and the other members of the Magnificent Seven. This way I would learn as much as possible about our North American shorebirds and get to know the rural nooks and crannies of the Lower 48, Canada, and Alaska. Using my diaries, field notes, more than 5,000 digital photographs, and my most vivid memories, I report on what I discovered in the narratives that follow. Those narratives are interwoven with technical discussions on the natural history of the shorebirds and boxed accounts describing the life histories of various of our North American shorebird species.

Before diving in, I need to provide additional definitions and reader guidelines. First, the term "North America" in this book denotes the Lower 48, Alaska, and Canada (it excludes Mexico, Central America, and Hawaii). This is the birder's definition, following the *National Geographic Birds of North America* as well as the *Sibley Guide to Birds*. Second, I use a narrow definition of the term "shorebird" for the taxonomic focus of this work. When discussing "the shorebirds," I am referring to the family Scolopacidae—the sandpipers and their close relatives—but excluding the plovers, stilts, avocets, and oystercatchers, groups that were included in Stout's famous *Shorebirds of North America*. The reasons I use "shorebird" for the Scolopacidae is simple: there is no available term restricted to the sandpipers and their close relatives aside from "sandpiper," but I do not feel "sandpiper" adequately covers the full breadth of the Scolopacidae. That said, in a few places I use the term "sandpiper family" interchangeably with the Scolopacidae. Hence, in this book Scolopacidae = shorebirds = sandpiper family.

Why am I focusing only on the sandpipers/Scolopacidae? Simple! It is because this bird family (to the exclusion of the others mentioned) encompasses the great global migrators. Their incredible travels are what inspire me and what I would like to present to the reader.

In this work, the species names of birds, other animals, and plants are capitalized following common ornithological practice. This allows readers to immediately discern the difference between the name of a particular species and generic description: a "Blue Jay" is the name of a species, whereas a "blue jay" could refer to one of several jays that are blue (such as Steller's Jay, Florida Scrub-Jay, Pinyon Jay, and of course the Blue Jay).

In the text boxes that encompass shorebird species accounts, I often discuss the conservation status of species as ranked by the International Union for Conservation of Nature (IUCN) Red List. This is a classification system of threat to the animal and plant species of the world. Rankings include, from lowest to highest threat: Least Concern, Vulnerable, Near Threatened, Endangered, Critically Endangered, and Extinct. As well, I occasionally mention the US North American Bird Conservation Initiative (NABCI) Watch List, which ranks all bird species of Mexico, the United States, and Canada by level of conservation threat. Those shorebird species considered threatened by NABCI will be mentioned as "on the Watch List." Note that the IUCN list is global and encompasses all plants and animals; the NABCI list is regional and focused only on birds.

Now let's dive into the story of North America's incredible shorebirds. ❋

PART I
WE GO WAY BACK

An adult male Dunlin in breeding plumage calls from his tundra nesting territory.

1 Birding Beginnings

Whirrrr! Up whistles a prime woodcock, giving a glimpse of russet plumage, long bill, and a flirt of white tail feathers as it tops out and dodges away over the saplings. It is as if a handful of fallen leaves had been whirled up by the wind.

—Henry Marion Hall, *A Gathering of Shorebirds*

I HAVE ALWAYS LOVED SHOREBIRDS and the places that attract them. Growing up in Baltimore, I did not have many opportunities to see shorebirds or visit their favored haunts. None bred in the city and only a few migrants would visit the city proper. As children, we would come upon shorebirds from time to time when exploring our favorite green spaces.

In the early 1960s, a patch of mature bottomland woods separated the campuses of Gilman School and Friends School near my home. Year-round, the neighborhood kids, me included, would traipse into the Gilman Woods in search of adventure. Stony Run ran through the lowest part of the woods, and a huge boulder marked the very middle of this urban wilderness. We called the boulder Tramp Rock. How this giant boulder got there remains a mystery,

A Spotted Sandpiper in breeding plumage with a Killdeer in the background.

since Baltimore lies well south of the glaciated East Coast. The stream and the adjacent big rock formed the epicenter of our adventures in the Gilman Woods. We played war games, threw rocks into the water, and occasionally picnicked atop the rock.

At this time I was already interested in birds, and my home environs were still quite birdy. Coveys of Northern Bobwhite quail and the occasional Ring-necked Pheasant passed through my tiny backyard. Migrating warblers, mainly Myrtles, stopped over in the Gilman Woods in numbers in spring, and Wood Thrushes sang in the damp leafy bottomlands. Through those childhood years, two shorebirds enlivened those springtime woods: the Spotted Sandpiper and the American Woodcock. These, then, were my neighborhood shorebirds. They were the ambassadors of their family, introducing me to this great worldwide lineage of birds—the Scolopacidae.

The Spotted Sandpiper—or "Spotty"—was the more familiar of the two. In late spring it was not uncommon to encounter a solitary individual along Stony Run, foraging among the stream rocks, its hindquarters teeter-tottering up and down as it hunted for stream- and gravel-loving invertebrates. The peppering of large black spots on the snowy-white breast made it easy to identify. When flushed, it would fly low over the stream with its stiffly fluttering wings held low. This bird's weird behavior was so unlike the songbirds in the woods; its body was always moving up and down, and it behaved like some kind of drunken windup toy. Spotties showed up mainly in the spring, when they displayed their handsome breeding plumage. In flight, the sandpiper would emit a series of rising musically piping notes, making it all the more adorable.

Even more unusual was an encounter with the American Woodcock. These oddball shorebirds stopped over in the leafy bottoms of springtime woods, well shaded by tall oaks, not in association with the creek. This is a shorebird that eschews the shore. And unlike a sandpiper, this bird stands low to the ground, has an abbreviated neck, a chubby, globular body, and an outsized bill used for hunting earthworms. The woodcock's plumage matches the dead leaves on the ground, making it difficult to detect in its woodland haunts. It is, no doubt, the least shorebird-like of the family.

My youthful encounters with the woodcock tended to be heart-stopping affairs. As I passed by a hidden bird, my footsteps would flush it from its hiding place on the shaded ground. The noisy explosion of wings and near-vertical takeoff of the bird would send my heart into my throat and make me jump back in surprise. It would take me a moment to realize that I had startled a woodcock, and I needed another minute to recover from the shock. This was always an event that merited a report back to my family at the end of the day. In these years, birds were becoming a part of my daily experience. I was growing into an active bird-watcher.

When I was a young and ambitious birder on the East Coast in the late 1960s, shorebirds topped my list of "most wanted" birds. There were rarities, like the curlews and the godwits, that were avian gold. Because these birds are great migrators, they tended to bypass my local birding haunts, only touching down in famous coastal reserves, distant from my home. Encounters with these rarities marked notable points in the passage of a year, or even decade, in my birding diary. The birding highlights of those days, in the 1960s and '70s, left a lasting impression on me and remain some of the sweetest memories of my more than a half-century in pursuit of shorebirds.

In 1969, the summer before my senior year of high school, my mother and I vacationed for a week with family friends in Hancock, Maine. Northeastern New England in August features long days of deep blue skies, soft breezes, and a rich migratory birdlife that is starting to move in anticipation of the approaching autumn. In Maine I had the chance to add to my small but growing life list of birds I had seen. I spent the vacation on the lookout for novel birds while also fishing on George's Pond, climbing Sargent Peak on nearby Mount Desert Island, and taking walks along the rocky Maine coast.

On one afternoon, three of us wandered along the Corea shoreline that faces the Sally Islands. At low tide, a wet sandflat led out to nearby Bar Island, which was only accessible because the water was at low ebb. Herring Gulls loitered on the hard-packed sand and soared overhead. Not far offshore, pairs of Black Guillemots bobbed about and dove for small fish. Clots of female and young Common Eiders in their drab plumage watched us cautiously from a distance.

We followed the sandspit out to Bar Island, enticed at being able to walk to an uninhabited Maine isle. Common Terns swooped overhead. A Least Sandpiper gingerly picked at seaweed piled up at the edge of the sand. Suddenly, our peaceful reverie was broken by the strident call of an unknown water bird: a rapid series of liquid piping notes. We had flushed a large shorebird that had been foraging among the algae-covered rocks fringing the island. It sailed low over the water to an adjacent islet, all the while calling out.

Through my decrepit 7 × 50 World War II vintage binoculars, I could see a shorebird that was unfamiliar to me. I signaled to my mother to put her binoculars on the bird. The shorebird appeared much a part of the scenery of somber grays, browns, and greens of the rocky Maine coast. It blended in with its surroundings—it was dark gray-brown above, finely patterned and paler below. But most distinctive was the head, with its blackish decurved bill, black eyebrow, and black eye stripe. And it stood on long blue-gray legs.

As we waded toward the bird, it flew off, its plaintive cries trailing behind. That single shorebird made my Maine holiday. It was what Peterson's *Field Guide to the Birds* at that time termed a "Hudsonian Curlew" and what Robbins's slightly newer *Birds of North America* termed a "Whimbrel." (I wrote "Whimbrel" in my diary that afternoon, so I must have been using the Robbins guide as a reference that day.)

The Whimbrel is North America's most widespread large shorebird and is one of the flagship species of the family Scolopacidae, the focus of this book. In the decades since that initial encounter, I have seen Whimbrels in a dozen spots across North America. But in 1969, I had been longing to see this handsome and uncommon bird during its twice-annual passage between its Arctic breeding ground and its main wintering grounds in Central and South America. The various memories of my encounters with the Whimbrel over the years constitute a chapter in my life with birds. In my mind, I can still wing back to that first Whimbrel, perched on that seaweed-encrusted rock, in an earlier century, an earlier life; that image in my head brings me a poignant joy, with allied memories of loved ones no longer with us who shared that day in the distinct Maine light of a very fine August afternoon.

During August and September 1970 I vacationed with my girlfriend, Nancy, and her family at West Falmouth on Cape Cod. Nancy's father, an enthusiastic birder, invited us to join him on a day-long birding tour of nearby Monomoy Island, led by a team from the Massachusetts Audubon Society. Our boat negotiated the tides and treacherous sandbars and, once on terra firma, our group of 15 birders spent seven glorious hours rumbling up and down the long

An adult Whimbrel in an East Coast Salt Hay marshland.

sand island in a ramshackle dune buggy looking for birds. It was the height of autumn migration, and this was ground zero for birds stopping to rest and feed before heading south over the Atlantic in their perilous migration to the tropics. The day's weather was cool and clear. The vegetation over most of the island was low and trees were few, mainly a scattering of dwarf pines and thick brush.

We recorded 60 bird species that day, including seven new "life birds" for me (species seen for the very first time), the rarest of which was a single Hudsonian Godwit, in its gray-brown autumn plumage, perched on a mudflat with a mixed flock of shorebirds. Godwits migrate in flocks. This was a stray individual that, for some reason, had stopped on this sand island rather than continuing its long overwater flight from Nova Scotia to Brazil. No matter, it was a magnificent creature, a rare thing—a little-known species living on the very edge of my young imagination. In his *Ornithological Biography*, John James Audubon wrote of this godwit: "This species, which is of rare occurrence in any part of the United States, is scarcely ever found farther south along the coast than the State of Maryland. I have never seen it in the flesh."

No birder ever forgets encountering a Hudsonian Godwit for the first time. It's one of those elusive, rare birds we dream about seeing. Bending over a spotting scope, I scrutinized the demure plumage of this shorebird, a large and long-legged sandpiper with a long bill that was slightly upturned. The plumage was a mix of gray and brown dorsally and pale buff below, and most distinctive was the cunning white eyebrow atop its dark eye.

The resting bird eyed us suspiciously and then arose and winged southward. When the bird lifted off, its striking jet-black underwings proved we had just encountered one of the rarest shorebirds in North America. As the bird disappeared, we all stood upright and eagerly eyed each other, assuring ourselves that this was not, in fact, a fever dream. It was encounters like this one that set the stage for adventures farther afield. What follows is the story of my grandest shorebird adventure of all, as well as the story of the shorebirds themselves. ❋

2 Natural History of Shorebirds

Here, where breakers thunder down and flying spray obscures the scene, the little Sanderlings, ever on the alert, run nimbly into the returning flood to snatch up many a choice tidbit, and then trip lightly up the slope, ahead of the incoming wave.

—Edward Howe Forbush, *Birds of Massachusetts and Other New England States*

ANYONE WHO HAS SPENT TIME on the sandy shores of the United States or Canada knows a sandpiper. Sandpipers form the core lineage that encompasses the avian family Scolopacidae. There is a "sandpiper look" that birders know—it's all about morphology and behavior. Two North American scolopacids—the Sanderling and the Greater Yellowlegs—typify characteristics of the sandpiper lineage, what we're calling "the shorebirds." The Sanderling is widespread along our seashores, common, and typical of many species in the family. It has a small, rounded head, with a slim elongated and pointed bill, plump body, pointed wings, longish legs, and three forward-pointing toes. Most of the time, the Sanderling is seen by beachgoers in its nonbreeding plumage, patterned dove gray above and mainly white below. In flight, the wings show a prominent white "flash" or wing stripe. The infrequently encountered breeding plumage features a rich rusty wash on the head, breast, and back. Birds do not put on this bright plumage until they move northward toward their Arctic breeding haunts in spring migration. As with many of the shorebirds, the Sanderling breeds in the Far North; migrating southward in autumn, some individuals winter as far south as Tierra del Fuego, off the southern tip of South America.

Larger and slimmer than a Sanderling, the Greater Yellowlegs possesses a slim and pointed bill, smallish eyes, a small head, and a slim and elongated neck. It has a slightly elongated body with long and pointed wings that protrude when folded to extend beyond the moderate-length tail, long legs, and a vestigial hind, or fourth, toe that is absent in the plovers and the Sanderling. Its pointed wings are distinctive when the bird is in the air, enabling its nimble and graceful flight. When foraging, it has a rapid gait without any of the cadenced hesitation exhibited by the plovers. The Sanderling typifies the smaller members of the group, whereas the Greater Yellowlegs is typical of the larger members of the Scolopacidae.

And what distinguishes a sandpiper from the other shorebird-like relatives—the plovers, oystercatchers, avocets, and stilts? The small, rounded head; the finely patterned dorsal plumage, often with dark vermiculation; the fine bill; the longish legs; and the swift aerodynamic flight. As summarized above, members of the sandpiper/shorebird family share morphological features, making each species readily recognizable as belonging to this family despite considerable diversity in form among the species. Many are long legged, whereas a few (like woodcocks and snipes) are short legged. Some are long billed, and others are short billed. But all exhibit the diagnostic horizontally cast body shape, the small head, and the shortish tail and longish wings. Most are seen around shorelines or mudflats, where they wade in search of animal prey. The diversity of bill shapes and lengths indicates a wide range of foraging styles and diets. Some, such as the Willet and Stilt Sandpiper, have partially webbed feet, a

Near its nest, a Greater Yellowlegs scolds an intruder from atop a White Spruce.

feature that presumably aids in their foraging when in deeper water. Most of the shorebirds, aside from the turnstones and Surfbird, have a slightly swollen tip of the bill that is filled with sensory cells that aid in detecting invertebrate animal prey buried in the mud or sand.

Nine Shorebird Lineages

The North American Scolopacidae family includes the woodcocks, snipes, dowitchers, godwits, curlews, shanks, turnstones, peeps, and phalaropes. The woodcocks (which include a single species breeding in North America) are the most atypical of the family, being woodland-dwelling, chubby bodied, short necked, short legged, plumaged in cryptic woodland colors, and having specialized wing feathers that produce a twittering sound when the bird carries out its springtime aerial display. The snipes (one species breeding in North America) are woodcock-like in most respects but are marsh-dwelling, streaked rather than barred, and have tail feathering that produces low-pitched hooting sounds in their flight display. The dowitchers (two species breeding in North America) are somewhat snipe-like but are Arctic and sub-Arctic nesting and flocking, and more closely resemble typical shorebirds. The godwits (three species breeding in North America) are tall, large, recurve-billed, wetland loving, and drillers into muddy wetlands when foraging. The curlews are tall, large, decurve-billed, sexually monochromatic, and pickers and drillers when foraging. The shanks (members of the genera *Tringa* and *Actitis*; six species breeding in North America) are graceful, slim billed, long legged, yellowlegs-like, and rather typical of the family. The turnstones (three North American breeders, including the Surfbird) are compact, short necked, short billed, Arctic breeding, and coast loving in winter. The peeps (13 species breeding regularly in North America), include those in the genus *Calidris* except for the Surfbird (which I place with the turnstones). The peeps are the largest lineage of the family; they are small or smallish, short billed, short legged, compact, and short necked. They are commonly seen in flocks. The three phalaropes (all breeding in North America) are atypical; females are larger and have brighter plumage, and males manage nesting duties. They are water loving and twirl when foraging in water. Two of the phalarope species spend most of each year out in the open ocean far from land.

Evolutionary History

The Scolopacidae or the sandpiper family (what I am calling the shorebirds) is a distinct evolutionary lineage of mainly shore- and wetland-frequenting birds that ranges across the globe, from the

High Arctic to southern Chile, South Africa, New Zealand, and Australia—only absent from Antarctica. The family as a group is seasonally mobile and many of the species migrate long distances. The family includes the woodcocks, snipes, dowitchers, godwits, curlews, shanks, Polynesian sandpipers, turnstones, peeps, and phalaropes—totaling 86 extant species globally and 55 species known from North America as breeders, migrant visitors, or rare vagrants (the list of "featured shorebirds" on page viii includes just those 35 species that are discussed substantially in this book—mainly birds I encountered during my solo travels).

The Scolopacidae is a family nested within the Charadriiformes, a large avian order composed of approximately 390 species of birds that also includes the buttonquails, coursers and pratincoles, auks, terns and gulls, jacanas, painted-snipes, seedsnipes, Crab-plover, Plains-wanderer, oystercatchers, avocets, Ibisbill, plovers, and thick-knees.

A 2022 study by systematists David Černý and Rossy Natale combined molecular, morphological, and fossil evidence to generate a phylogenetic tree that indicates that the Charadriiformes arose in the Paleocene epoch (>60 million years ago) and radiated strongly in the late Eocene and Oligocene epochs (45–28 million years ago). Thus our sandpiper lineages all evolved after the demise of the dinosaurs in the end-Cretaceous period. Scolopacid fossils that may refer to the godwits (*Limosa*) and shanks (*Tringa*) are known from the late Eocene. The best-characterized scolopacid fossils are of no earlier than the Miocene (which began 23 million years ago). It is interesting to note that this lineage, dominated today by Arctic tundra breeders, evolved mainly during a period when the entire planet (including the polar regions) was very warm. Therefore, tundra breeding by the shorebirds must be a relatively recent phenomenon. Where did this lineage originally evolve?

The 2022 systematic study postulated that the sister group (the evolutionarily closest lineage) to the sandpiper family is a large lineage that includes the jacanas, painted-snipes, seedsnipes, and the Plains-wanderer. It is interesting to note that all these listed "sister" families have a tropical or Southern Hemisphere distribution today. Perhaps the Scolopacidae constitutes the northern counterpart to this southern lineage, with the two evolving in separate hemispheres.

Morphology, Anatomy, and Physiology

The most striking aspect of sandpiper anatomy is the variability of the bill from species to species and from genus to genus. This is because the bill is pivotal to foraging specialization. Variation in bill morphology makes possible all manner of foraging styles, creating new ecological niches that allow for adaptive radiations of species.

For some, the bill tips are rounded and slightly bulbous to hold many specialized cells that populate receptor organs such as Herbst and Grandry corpuscles, which link to parts of the brain that engage in tactile sensing of the bill tip. These specializations are absent from the bills of the plovers. Such receptors presumably allow the shorebirds to harvest prey efficiently even when the bill tip is underwater or deep in the mud or sand. The tips of the upper and lower mandibles of some species are mobile and flexible (exhibiting rhynchokinesis) and help the birds capture and manipulate small, slippery, living prey items.

Some shorebirds (e.g., Dunlin and Western Sandpiper) possess novel adaptations to their bill and tongue that allow them to graze on nutrient-containing biofilms that develop on the underwater surface of intertidal flats. The birds possess denticles on the roof of the upper mandible that enable the bird to scrape the food-carrying biofilm material from the tongue, which itself captures the biofilm in water using mucus-encrusted papillae that are coated with microvilli.

The guts of shorebirds are rather uniform, featuring from front to back an esophagus, a glandular stomach (known as the proventriculus), the muscular stomach (gizzard), and then the small and large intestines. The large intestine features two small caeca, perhaps for digestion of fibrous matter. Those species (e.g., Red Knot) that forage on mollusks exhibit a large gizzard for breaking up the shells of those prey that they swallow whole. Many shorebirds eject from the mouth inedible material, such as shell fragments, as pellets. More specialized mollusk-feeders pass the fragmented shell bits through the large intestine. The gizzards of shorebirds vary in size with the season, based on functional need. Whimbrels are known to seasonally shed the lining of the gizzard, hypothesized to rid individuals of a buildup of parasites.

Finally, preparatory to the long autumn migratory flights over water, digestive organs may grow large to aid the accumulation of fat and muscle and then rapidly and substantially shrink immediately prior to migration, to be converted into stored fat and muscle to power prolonged flight. Whereas shorebirds need to ingest about a third of their body weight daily to subsist, prior to long migrations they must greatly expand their food intake.

Shorebird physiology is complex and closely allied to each species' annual migration demands and breeding requirements. Male sandpipers heading north to the breeding grounds see their reproductive organs develop in advance of arrival, whereas those of the females do not fully develop until after the birds arrive in the north. Testis size of males relates to mating system. Polygynous males carry larger testes than monogamous species; the males who will be mating with multiple females will need more sperm. The lek-breeding male Ruff (a Eurasian species) has the largest testes of all the scolopacids. The sperm cells of scolopacids are uniquely spiral in form, which may relate to high levels of sperm competition (in which the sperm of various individual males that mated with a single female compete internally to fertilize that female's eggs).

Marine-dwelling species of sandpipers have salt glands in their foreheads to help excrete excess salt from the blood plasma. Red Knots, which consume whole mollusks from marine environments, exhibit the largest salt glands of any sandpiper. These salt glands enlarge and regress depending upon the birds' current environment. Land-based shorebirds such as the snipes and woodcocks lack these organs. Wilson's Phalaropes spend much of their lives in saline habitats but exhibit smaller salt glands because they can rid their prey of salt before consumption.

For super-migrators such as godwits and some curlews—bird species known to migrate very long distances—the entire digestive apparatus (including liver and kidneys) enlarges during the staging period to aid the buildup of fat stores and then becomes much reduced just prior to takeoff, transferring those nutrients to help grow the heart and breast muscle to aid flight efficiency. This is an astonishing short-term anatomical turnaround that permits those long overwater flights for which these birds are justly famous. The super-migrator shorebirds depend on massive breast musculature for their extensive spring and fall travels. These species possess a strongly keeled sternum, which features a high ridge down the center to which flight muscles are attached. The pectoralis muscle powers the downstroke and the supracoracoideus powers the upstroke of each wingbeat. These breast muscles are the largest component of the shorebird's nonfat mass.

Wing shape in the shorebirds varies considerably. Super-migrators possess long and pointed wings that are good for high speed and high-altitude flight over long distances. The snipe and woodcock, which are ground-nesting woodland-dwellers with reduced migration demands, have shorter and more rounded wings; these wings are most useful during the nesting season, when there is need to escape from predators using an explosive vertical takeoff from the ground. Such wing shapes cannot be much benefit during seasonal migration. The Spotted Sandpiper has rounded wings and a weird, hovering flight low over the water, with the ability to drop into the water to escape from an attacking predator. This species makes a migration of short hops, quite distinct from those of the super-migrators.

Spotted Sandpiper ***Actitis macularius***

Other names: Spotty, Tip-up, Tip-tail, Land-bird, Teeter, Teeter-bob, Teeter-peep, Twitchet, Peet-weet, Teeter-snipe

Appearance: One of the most familiar of the shorebirds, the Spotted Sandpiper in its breeding plumage is strikingly marked with distinctive black spots on a white breast and has a mainly orange bill. Wintering birds are plain-breasted and dark billed.

Range: The Spotted Sandpiper breeds widely across the northern two-thirds of the Lower 48 and most of Canada and Alaska. It winters from the southern United States to Argentina.

Habitat: In spring migration, the Spotted Sandpiper appears as singletons in many places where there is water for foraging—streamsides, ponds, and ephemeral pools in the woods. Autumn migrants and wintering birds are sometimes found in a wider array of habitats, including sandy beaches, lagoons, and mangroves.

Diet: The bird's diet includes small crayfish, fish, mollusks, worms, crabs, and insects. The bird forages by stalking and picking at the surface of the water and ground for prey items, and it may rush and leap after flying arthropods.

Behavior: The larger female will mate with as many as four males in a season. Both males and females conduct low display flights with a mix of flapping and gliding, wings held low and head held high, while vocalizing. Females arrive on the breeding ground first and defend a nesting territory to attract a mate. After breeding, adults start moving south in early July, the young in August. Some winter in the Lower 48, but others migrate to the Southern Hemisphere.

Nesting: The nest is a scrape on the ground near water or sometimes under a bush. Females lay four creamy-buff eggs, thickly spotted and speckled with dark brown. The male raises the young.

Conservation: The species has declined by as much as 54 percent since 1980, and yet the species remains one of the most widespread and common shorebirds of North America, totaling 660,000 individuals. It is classed as Least Concern by the IUCN.

Differences Between the Sexes and the Ages

Some sandpipers exhibit strong sexual size dimorphism. For instance, the male Ruff weighs one and a half times that of the female (known as the Reeve). The polygynous breeding Pectoral Sandpiper male is also considerably larger than the female. By contrast, in four other lineages (godwits, curlews, knots, and phalaropes) the females are larger than the males. Finally, in the woodcocks, snipes, and shanks, the males and females are of equal size.

Plumage dichromatism (differences in color) between the sexes is pronounced in the phalaropes, in which the polyandrous females are brighter colored. In most godwits, the breeding males exhibit a substantially brighter plumage than the females. The maximum plumage dichromatism is exhibited in the Ruff (which, recall, is a polygynous lek-breeder). The adult female is dun colored and plain, but the breeding males grow a large strikingly colored ruff of feathers

An adult Pectoral Sandpiper forages in its favored migratory habitat.

about the face and throat that varies from male to male (analogous to the male Lion's impressive mane). Male Ruffs in the breeding season sport these gaudy plumes to attract females for mating.

Breeding male Ruddy Turnstones show diagnostic facial patterns of black and white that enable researchers to identify individual birds year after year, and which apparently allow males to recognize each other and perhaps make it possible for females to recognize potential mates. This might have some advantages in mate competition and acquisition.

Shorebirds exhibit three distinct plumages related to age and season: adult breeding, adult nonbreeding, and juvenile. Typically, the breeding plumage is the brightest and prettiest, with colorful highlights (e.g., the Stilt Sandpiper with its rusty eyebrow and the Western Sandpiper with its rufous shoulder, cheek, and crown). The nonbreeding plumage is the dullest, plainest, and palest. Juvenile plumage tends to look like nonbreeding plumage but with the mantle heavily marked because of pale edgings to the dark mantle contour feathering. This last plumage grows in before the bird departs the nesting grounds and is preparatory to the individual's first winter. One sees many birds in juvenile plumage in autumn staging and wintering sites in the Lower 48.

Dietary Diversity

Shorebird diets vary in substantial ways, not only from species to species but also from habitat to habitat and season to season. Most shorebirds seek very small prey, but the range taken can be considerable. Red Knots and Ruddy Turnstones feast seasonally on the tiny spherical eggs of the Atlantic Horseshoe Crab in Delaware Bay. Many tundra-breeding shorebirds forage mainly on small arthropods and their larvae. Wintering and staging birds probe soft substrates (often underwater) to collect mollusks and polychaete worms (segmented worms known informally as bristle worms), while others, such as yellowlegs, will chase after small fish in shallow water.

Small sandpipers feed most successfully with the receding tide. And don't forget the fiddler crabs that are loved by the Whimbrel. Or the American Woodcock sucking up earthworms on the floor of the forest interior. Or the Red Phalarope bobbing on the surface of the stormy Atlantic for months on end, harvesting planktonic marine microfauna from the sea's surface.

The Ruddy Turnstone, with its habit of overturning all manner of beach riprap in search of edible treats, enjoys a breathtakingly broad diet, ranging from gull excrement to bits of whale carcass. That said, the turnstone does show a preference for crustaceans and fish. This turnstone and the Bristle-thighed Curlew are known to steal and consume eggs of colonial nesting birds. Some individuals have even been seen to take the thick-shelled eggs of nesting albatrosses; to break these open, the curlew will hit the egg with a piece of coral rock—a unique example of tool use by a shorebird.

The three species of phalaropes are known to spin while floating in the water. Apparently, this brings tiny prey items to the surface, and these prey items are then drawn into the bill by surface tension. In autumn, huge flocks of Red-necked Phalaropes gather on the Great Salt Lake of Utah to feast on the hyperabundant brine shrimp and brine flies there (338,000 were counted there in August 2019). Wilson's Phalaropes can also snap up tiny flying invertebrate prey. Peeps probe soft sediments to harvest amphipods (shrimp-like crustaceans with a laterally compressed body) and various marine and aquatic worms. The godwits subsist on small bivalves, large polychaete worms, and diminutive crabs. Bivalves and gastropods harvested from rocky substrates along the coast are favored by Purple Sandpipers, Surfbirds, and Wandering Tattlers in winter. Sanderlings seem to prefer small mole crabs that live buried in the wave-washed sand. These are also taken by Red Knots. Dowitchers appear to target worms of the family Nereidae.

Most shorebirds subsist mainly on animal matter. By contrast, Hudsonian Godwits feed on aquatic vegetation found in freshwater ponds in Canada, both during migration and on the breeding grounds. Right after arriving on their tundra breeding grounds, Red Knots have been observed to take small seeds, as well as young shoots of horsetails. In autumn, southbound Eskimo Curlews (which are now extinct) were known to consume quantities of crowberries in Atlantic Canada before setting out for South America. Newly hatched scolopacid chicks feed themselves by foraging around the nest, taking mainly spiders and emerging arthropods. Prebreeding females and young chicks require additional calcium to build eggs and bones, respectively; these birds consume lemming teeth and bones, something not done by the adult males.

What of the Long-billed Curlew and Marbled Godwit and Upland Sandpiper in their high plains redoubt? They mainly hunt for arthropods. They stalk the grasslands and plains, hunting for moving invertebrates—grasshoppers, crickets, beetles, beetle larvae, caterpillars, spiders, centipedes, and even snails. All these birds also ingest a small quantity of plant matter (seeds, small berries, and roots), whether on purpose or simply from the habit of picking and swallowing anything that looks edible. Long-billed Curlews have even been recorded taking the eggs of other birds as well as toads. For the godwit and curlew, the winter diet shifts to marine invertebrates, whereas for the Upland Sandpiper, the arthropod diet continues to dominate in its wintering habitat on the South American Pampas.

Habitat Selection

We generally think of shorebirds as skirting the seashore or foraging on mudflats and sandflats, mainly in association with brackish or salt water. Certainly these are very popular habitats for many species of sandpipers. But given the mobility of these birds and their migratory habits, we can find them in a broad array of both coastal and interior environments, depending on the season and their travels.

Spring brings a flood of migrant shorebirds northward. Northbound Short-billed Dowitchers and Western and Semipalmated

Sandpipers like to forage on muddy coastal flats. On the other hand, Red Knots prefer to feed on sandy intertidal flats. Many head to specific favored staging sites, such as Reeds Beach on the Delaware Bay in New Jersey, to fatten up on the eggs of the Atlantic Horseshoe Crab. Hudsonian Godwits stop over in Cheyenne Bottoms, Kansas, to harvest freshwater polychaetes. Favored habitats include productive shores, mudflats, sandflats, shallow estuaries, expanses of short-mowed grass, flooded bare-earth agricultural fields, sewage treatment plants, and even sod farms. In spring, northward-bound Red Knots and Hudsonian Godwits stop to forage in shallow coastal lagoons in southern Brazil.

A few shorebirds breed in the Lower 48: the Spotted Sandpiper in wooded streamsides, the American Woodcock in bottomland forests, the Upland Sandpiper and Long-billed Curlew in short-grass prairie, and the Marbled Godwit in wet swale grasslands. Most shorebirds, however, head up to northern Canada and Alaska, where they make their nests in bogs, openings in spruce forests, and various tundra habitats. In fact, more species choose tundra over any other breeding habitat.

Early summer sees the return to the Lower 48 of those birds from the Far North that failed to snag a mate or a territory. They head down to popular loafing sites along both coasts and to some favored interior wetlands. These birds can appear as early as the end of June along the shore of New Jersey, even though their successful counterparts may not start moving south for another month.

Fall migration starts in earnest in August, when the successful breeders head to staging sites in Saskatchewan and James Bay, where the foraging is good and they can build up large reserves of fat for their long flights to South America. Whimbrels fatten up on the sociable fiddler crabs that build burrows at the highest parts of the coastal intertidal zone; they also feed on a variety of small berries (such as blueberries and crowberries) along interior grassy swales.

Winter sees most of our North American breeders moving to subtropical, tropical, and south temperate climes. These birds find similar coastal and estuarine flats in Argentina and Chile, where they can spend the months of the northern winter in a summer clime south of the equator. It is here one can find huge monospecific flocks wintering in South America's very richest coastal feeding sites for these birds.

Surfbirds, Purple and Rock Sandpipers, and both turnstones settle down for the winter on rocky coastal shores of Canada and the Lower 48, where they subsist on marine invertebrates that cling to the rocks and seaweed washed by the beating waves of the Atlantic or Pacific. Small numbers of lingering Whimbrels and Marbled Godwits haunt the backwater estuaries where marsh grasses give way to mudbanks rich with other invertebrate delectables. They will be joined by large Dunlin flocks and the occasional yellowlegs and Least Sandpipers—and the ever-elusive Wilson's Snipe.

It is noteworthy how much habitat selection can shift from breeding to wintering. Witness the Red Knot: in summer it picks at the tundra to collect arthropods and their larvae in a dry or freshwater Arctic landscape; in winter it probes for mollusks on saltwater coastal flats. The Solitary Sandpiper, in contrast, remains faithful to its wooded streamside or pond both summer and winter.

Some species are quite habitat specific. For most of the year the Short-billed Dowitcher inhabits and forages in saltwater habitats, whereas the Long-billed Dowitcher prefers freshwater habitats and avoids salt water. Habitat choice is one of the best ways to distinguish these two nearly identical species.

Mating Behavior

Long-distance migrants may begin to vocalize and start mating displays at stopover sites far south of the breeding zone. For instance, in my travels, I saw male Buff-breasted Sandpipers performing their open-wing courtship display to females in fields in Nebraska weeks

An adult Solitary Sandpiper in its streamside feeding habitat.

before they were due to arrive at their Arctic breeding grounds. Other shorebirds that display midcontinent include Least and Western Sandpipers, Short-billed Dowitchers, Dunlins, and Red Knots. But true pairing of individuals to breed does not take place until the birds arrive at their breeding destinations.

Most shorebirds deploy a monogamous mating strategy in which the male and female form a pair bond at the breeding site, and both parents sit on the eggs and care for the offspring. This is the case for the Hudsonian Godwit and Long-billed Curlew, for instance. Males of most monogamous species perform vocal display flights over their territories. Variations of monogamy exist. In the simplest and best-known form, a single male and female team up to produce a clutch of young in a season. Female Sanderlings do a bit more. They will lay a clutch with a male, let him brood that clutch, and then take up with a second male and produce a second clutch for that male. This is called serial monogamy. The female engages with one male at a time but over the season is polyandrous (the term that describes a female that mates with multiple males).

Some shorebirds exhibit polygynous mating, in which the male mates with multiple females. This is exemplified by the lek-breeding Ruff; males of this species compete among themselves for favored mating sites, and successful males provide the female with nothing more than a donation of sperm. The male Ruffs exhibit at a classic clustered lek, usually situated in a short-grass pasture (mainly in northern Eurasia). As many as a dozen males will attend the lek during the breeding season. Females visit the lek to examine the males and select a mate. Ruff males come in two "flavors"—*independent*, sporting black or rusty-brown ruff feathers, and *satellite*, sporting a white ruff. The independent males hold mating territories, whereas the satellites float about and visit territories of one or more of the independents. The physically smaller satellite males occasionally get to "steal" copulations by their associations with court-holding males. It has been shown that the two behavioral classes of males are determined by genetics. Individuals cannot switch mating type. Arctic-breeding Buff-breasted Sandpipers also exhibit a lek mating system, but the males do not feature any of the extravagant plumage specializations found in the Ruff. In all lek systems, the female raises the offspring without assistance.

When males are in excess, phalaropes exhibit polyandry; the colorful and larger female mates with multiple males, leaving the males to brood and care for the multiple clutches she lays. This differs from the serial monogamy of the Sanderling. In the phalarope, the female will mate with a number of males and produce clutches for several of these, which do all the brooding and parental care.

Shorebirds mate on the ground or, in the case of phalaropes, in the water. For those species in which females mate with multiple males, there is evidence of sperm competition. Females of certain

American Woodcock *Scolopax minor*
Other names: Little Whistler, Bogsucker, Timber Doodle, Hill Partridge, Wood Snipe

For naturalists, the annual pilgrimage to experience the evening display flight of the woodcock is one of the joys of spring. Most other encounters involve flushing a foraging bird from the woodland floor. Under less fraught circumstances, one might observe a woodcock on the leafy forest floor as it probes for earthworms.

Appearance: This solitary inhabitant of our moist woodland openings is our strangest-looking shorebird. The bird has the look of a chubby brown ball of feathers with no neck, a blocky head, and a long, straight bill that the bird drills into the ground. The dorsal plumage is mottled dark and pale in a dried leaf pattern to make the bird virtually invisible when frozen in place on the ground.

Range: The woodcock breeds across eastern North America, from Newfoundland and southern Manitoba to northern Florida and northeastern Texas, nesting in overgrown fields, woodlots, and bogs. The species winters in the southeastern United States, from Oklahoma and New Jersey to Texas and Florida.

Habitat: In recent decades, national, state, and local wildlife organizations have initiated programs to craft clearings in blocks of forest, which are favored by the birds for nesting habitat.

Diet: The species eats earthworms and buried arthropod larvae.

Behavior: The Woodcock never mixes with other shorebirds. Big autumn flights of woodcocks occur during the passage of cold fronts. It is an early spring migrant.

Nesting: The ground nest is lined with leaves, and the four eggs are pinkish, blotched with brown and gray. The female builds the nest and raises the offspring on her own.

Conservation: This is a species of conservation concern and is included on the NABCI Watch List. Some 350,000 birds are bagged a year in autumn hunts in North America.

polyandrous species are known to store viable sperm from multiple individuals for an extended period before the eggs are fertilized (known as sperm storage). This phenomenon allows females to conceal which males were successful in fertilizing eggs, which encourages all the males to invest time and effort in assisting at the nest. This constitutes true polyandry.

Sound Production

The American Woodcock carries out its famous display flights high overhead in wet clearings on spring evenings. The bird's wings produce twittering sounds during the display, all generated by wind rushing through three specially modified outer primary feathers. Another lek-breeder, the Pectoral Sandpiper, carries out a low song flight over the tundra, during which it inflates its breast sac, which produces a rhythmic low hooting that speeds up with time. Interestingly, two of the most famous lek-breeders, the Ruff and Buff-breasted Sandpiper, allow their postures to dominate the display.

Contact calls are commonplace when individuals are forming a flock to initiate a long-distance migratory flight. Individuals continue to vocalize once the flock is formed, and their voices may be heard periodically throughout the flight. Yet other vocalizations relate to pair formation and premating communication, as well as to warning fledglings of an approaching predator or to gathering up fledglings to warm them under the parent.

Nesting

Virtually all shorebirds place their nest on the ground, although a few of the shanks lay their eggs in the discarded nest of a songbird situated up in a conifer. The most common nest is a vegetation-lined depression or scrape into which the eggs are laid. Some species, such as Baird's Sandpiper, produce a neat cup nest on the tundra. Surfbirds nest high on a ridge among rocks. The Wandering Tattler nests on a riverine gravel bar. The American Woodcock situates its ground nest among leaves in the shaded woodland interior.

No shorebird lays more than four eggs, but some lay as few as two. Most species lay four. The nesting birds depend on crypticity while sitting on the nest and tend not to flush when a potential predator passes by. The patterned dorsal plumage of the sitting adult often matches the surrounding tundra vegetation, making the detection of the bird on the nest nearly impossible. The parents start to incubate the eggs after the entire clutch has been laid. Both parents carry out incubation duties among the monogamous species, though in many species, the female departs after the eggs hatch and the males are left to look after the precocial young (which are well developed and able to walk and feed themselves). Male care includes brooding the young during cold and rain, as well as warning them when a predator approaches. This tendency by the female to depart the nest indicates that these shorebirds do not maintain long-term pair bonds and that pairs apparently do not migrate together. Only the snipe and woodcock adults provision their young offspring. Quite remarkably, in all other species the hatched nestlings feed themselves from day one.

Migration

Most shorebirds nest in the northern regions (Arctic, sub-Arctic, and boreal zones) and spend the winter far south of their breeding habitat. For instance, the White-rumped Sandpiper population that breeds in Arctic Canada winters mainly in southernmost South America. In spring, the wintering birds make their way north in a series of flights, leapfrogging to their breeding grounds, only stopping to feed and rest. For birds in the Southern Hemisphere, this can begin as early as February. Because of the wind systems, these northbound birds tend to arrive on the coast of Texas and then make their way up the middle of North America, stopping in the northern Great Plains, including central Saskatchewan (to which they will return in their late summer southbound travels), and then they move to their breeding site in western Alaska, where they will spend six or seven

A juvenile White-rumped Sandpiper wades in a Pantanal wetland (*foreground*) with White-cheeked Pintails and Roseate Spoonbills (*background*).

weeks producing the next generation of White-rumps. Those that fail to mate head southward to loafing sites (where nonbreeders loaf and hang out for the "lost season"). Some nonbreeding birds begin their southbound movement at the very start of the summer. At summer's end, pulses of birds head to staging sites where they feed and fatten up for their main southbound migration flight to South America. One of the best-known autumn staging sites for the eastern populations of the Hudsonian Godwit is the western shore of James Bay, Canada, where thousands may spend weeks feeding and resting to prepare for their big flight south.

The major migrators of the group fatten up in autumn in preparation for their long southward flights to the tropics or Southern Hemisphere. This makes these birds good eating. During the late nineteenth century, market hunters along the East Coast targeted the larger shorebirds, decimating their populations. Whimbrels, Eskimo Curlews, and Hudsonian and Marbled Godwits were shot for restaurant fare and the home table. Places that today are favorite autumn birding spots were once killing fields. Most of the shorebird species suffered, and the now-extinct Eskimo Curlew never recovered from this depredation. Others slowly recovered after passage of hunting laws in the United States and Canada in the early part of the twentieth century that banned the shooting of most shorebirds.

Epic Travels

Shorebirds that nest far north and winter far south perform Olympian feats of aerial mastery and navigation. Twice a year certain shorebird species make a journey of as many as 8,000 miles, often flying day after day over featureless open ocean. My beloved Hudsonian Godwits, which nest in Alaska, the Northwest Territories, and Manitoba, travel across North America and then out across the Atlantic to Amazonia and thence, after a rest, southward to southern Chile and Argentina. In the spring they carry out a return trip, crossing a section of the eastern Pacific on their way to the US Great Plains and then back to Canada and Alaska. Birds that can live as many as 20 years will travel in excess of 320,000 miles in their lifetime. Black-tailed Godwits have been shown to migrate, at times, at altitudes as high as 20,000 feet to find winds and air temperatures that benefit long-distance travel. Scientists don't yet understand how these birds can function in these extreme hypoxic environments.

Some shorebird individuals have been observed to start their long northbound migration on the very same day each year. What sort of time-keeping device informs the bird of the precise date? The more closely we examine the travels of these super-migrators, the more challenging it becomes to explain their extraordinary capacities. Brief accounts of the epic migrations of four shorebird species (Pectoral Sandpiper, Red Knot, Hudsonian Godwit, and Bar-tailed Godwit) follow. The mechanics of these improbable travels are discussed in Chapter 12.

The Pectoral Sandpiper winters mainly in South America, with small numbers traveling to Australasia. The northernmost breeding grounds are found on Melville and Bathurst Islands of Arctic Canada (latitude 76.5° north). In July, most Pectorals in northern Siberia head east, some apparently over the Arctic Ocean to Alaska and Arctic Canada before turning southeast and crossing the Lower 48 and going onward to South America. The North American breeders migrate south through the heart of the Lower 48. In autumn, southbound Pectorals stop in the lower Mississippi valley to reprovision. They stay an average of 10 days at a particular foraging site, and in some instances extend this reprovisioning at a second foraging site in the valley. These birds winter primarily on the Pampas of South America. The annual migratory round trip for the most accomplished individuals of this species is a remarkable 18,000 miles, making the Pectoral Sandpiper one of the true super-migrators.

The Red Knot summers in the High Arctic of Siberia, Alaska, Canada, and Greenland, and winters in New Zealand, Australia, southern Africa, southernmost Argentina and Chile, Florida, and South Texas. Its maximal annual south–north migration (approximately 9,500 miles) exceeds that of the Pectoral Sandpiper because some knot populations breed farther north in Arctic Canada (on Ellesmere Island, at latitude 82° north). The longest overwater flight of the North American birds is 3,500 miles, from Atlantic Canada to

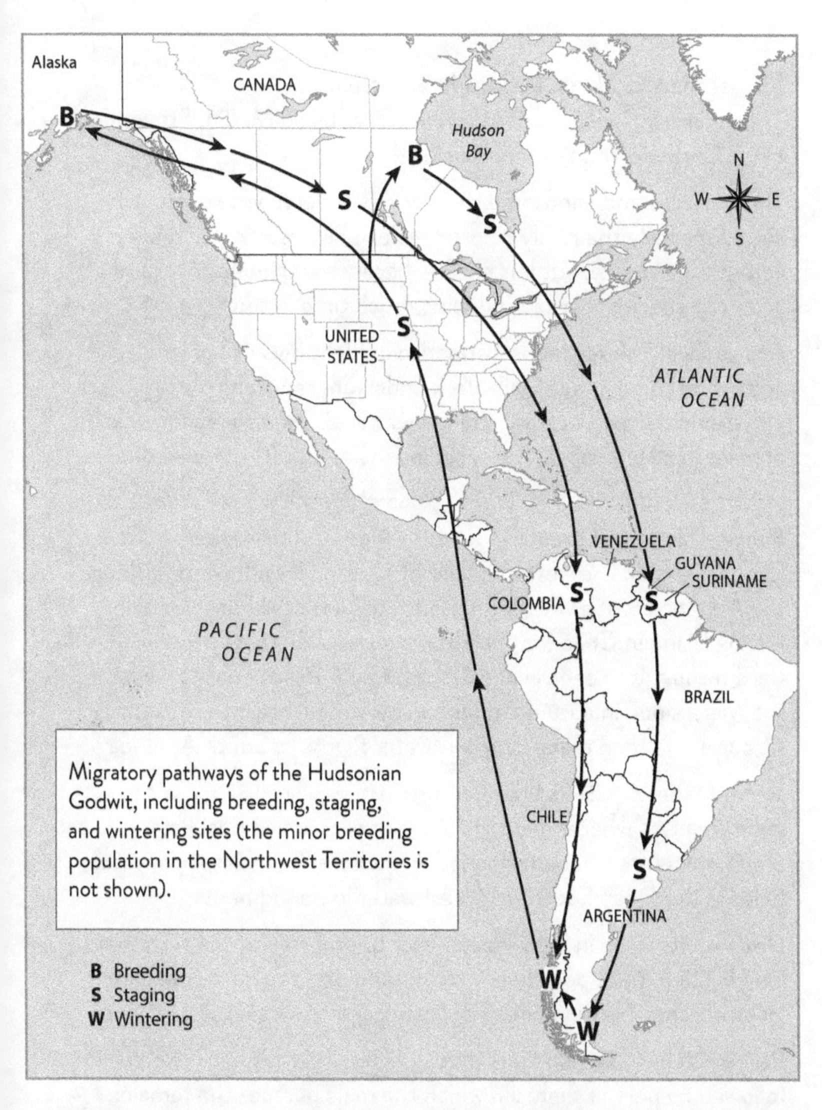

Migratory pathways of the Hudsonian Godwit, including breeding, staging, and wintering sites (the minor breeding population in the Northwest Territories is not shown).

northern South America. There the birds stop, rest, and feed before heading farther south in a series of leaps. In spring, many birds of the *rufa* subspecies stage in Delaware Bay and near the Nelson River in Hudson Bay before moving on to their breeding sites in the Arctic. One color-marked Red Knot named B-95 may have traveled more

Hudsonian Godwit *Limosa haemastica*

Other names: Red-breasted Godwit, Ring-tailed Marlin, Straight-billed Curlew, White-rump, Black-tail

This is a large and handsome shorebird that keen birders long to see—especially the richly colored male in his breeding plumage. Today it is not rare, but it is elusive, because long migratory jumps allow the species to avoid spending much time in the Lower 48.

Appearance: The species is distinctive in flight because of its black-and-white rump and tail, all-black underwing, and white wing stripe. The plainer female is larger and longer billed. The pale eyebrow is present in all plumages. The breeding male, with its marbled deep rufous underparts and striking, orange-based bill, is stunning.

Range: This godwit breeds in western Alaska, northwesternmost Canada, and on the western shore of Hudson Bay. It winters along the Atlantic coast of Argentina south to Tierra del Fuego and on Chiloé Island in Chile (on the Pacific coast). Migrants pass northward through the central and eastern Great Plains, mainly west of the Mississippi. In autumn, migrants move southeast to the Atlantic coast and then make long overwater flights to South America.

Habitat: These godwits breed in spruce bogs, usually adjacent to coastal waters where they can forage at productive tidal flats. Spring migrants frequent flooded agricultural fields (especially rice fields in the Deep South) and freshwater impoundments.

Diet: Adults wade in belly-deep water to probe soft mud with their long bill to extract polychaete worms and other marine or aquatic invertebrates. Nesting birds take many arthropods.

Behavior: The male conducts display flights over his nesting bog, followed by pursuit flights in which the male pursues the female.

Nesting: The nest is situated in a mossy hummock and features four deep olive eggs spotted with dark brown.

Conservation: The species is on the NABCI Watch List, with an estimated population of 77,000 that is in substantial decline.

than 430,000 miles over a 22-year period—farther than the distance from the Earth to the Moon, making the species a super-migrator.

The Hudsonian Godwit is a North American super-migrator that annually carries out a triangular migration across North and South America (see map on page 39). In autumn, birds that have been breding in western Alaska stage for a month in the Quill Lakes of Saskatchewan before making a prodigious flight southeast to the Atlantic coast and then southward out over the Atlantic to northern South America, where they stop in the Amazon basin for several days. They then jump to the Rio Gallegos estuary in Patagonia (Argentina) to feed on clams, mussels, and polychaetes. These birds then either fly southwestward to Chiloé Island in southern Chile or due south to Argentine Tierra del Fuego for six months of the boreal winter. Southbound, their longest known nonstop flight is 3,900 miles (from Hudson Bay to Venezuela, crossing 2,500 miles of the western Atlantic). Northbound birds travel nonstop from southern South America out across the eastern Pacific, crossing Central America to the Gulf of Mexico and touching down in the northern Great Plains to refuel and rest for several days. Their northbound traverse of the Pacific to Nebraska covers nearly 6,000 miles and requires seven to eight days of continuous flight. Getting from the Great Plains to western Alaska requires a final nonstop flight of several days traversing 3,500 miles.

The Bar-tailed Godwit of the subspecies *baueri* is the world champion super-migrator. Every autumn it crosses vast stretches of the Pacific, often in a single colossal flight. Its annual migration follows a triangular route. In autumn, the birds travel south-southwestward to New Zealand or Tasmania. In spring, the birds travel northwestward to the vast but disappearing mudflats of the Yellow Sea on the coasts of China and the Koreas. The godwits stage there for more than a month, then fly northeastward to western Alaska or eastern Siberia. This godwit's yearly travels require three distinct multiday overwater flights. The entire annual route covers 18,700 miles (see the map on page 42).

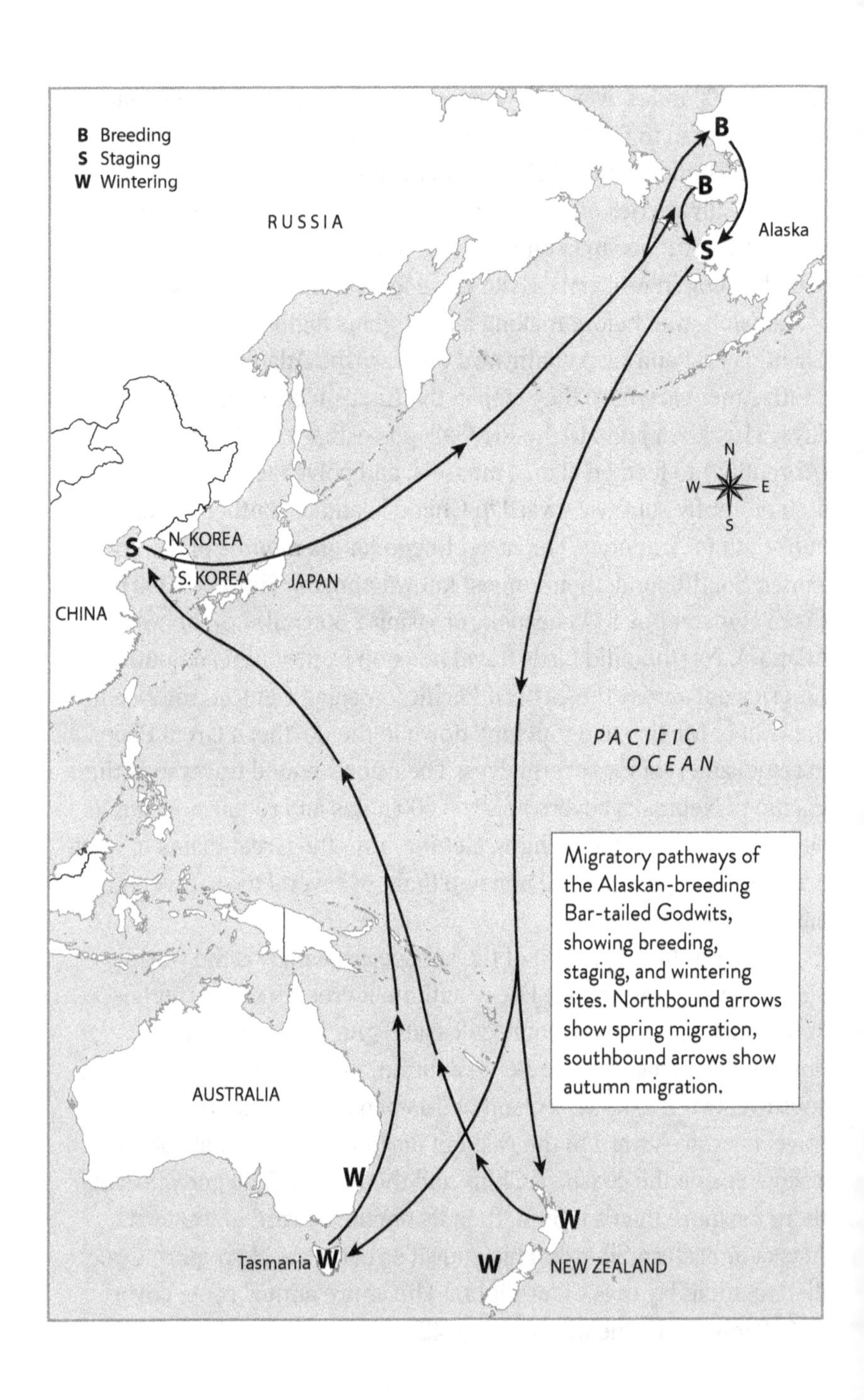
B Breeding
S Staging
W Wintering
RUSSIA
B
B
S
Alaska
N
W
E
S
S
N. KOREA
S. KOREA
JAPAN
CHINA
PACIFIC
OCEAN
Migratory pathways of the Alaskan-breeding Bar-tailed Godwits, showing breeding, staging, and wintering sites. Northbound arrows show spring migration, southbound arrows show autumn migration.
AUSTRALIA
W
W
Tasmania W
W
NEW ZEALAND

Bar-tailed Godwit ***Limosa lapponica***
Other names: Pacific Godwit

The Bar-tailed Godwit is one of the least accessible of North America's breeding shorebirds.

Appearance: The breeding male Bar-tailed Godwit is a handsome, rusty-breasted, large shorebird, not unlike the Hudsonian Godwit. The female is pale breasted with a bill longer than the male's.

Range: This widespread Arctic breeder nests from northern Scandinavia east to Siberia and across the Bering Strait to coastal northernmost Alaska and southeast to the Seward Peninsula and the Yukon-Kuskokwim Delta. Alaskan birds winter to New Zealand and Australia as well as Southeast Asia.

Habitat: Birders can find this species most easily in coastal Australia during the northern winter or in lowland and hilly tundra at Nome, Alaska, in the breeding season.

Diet: The species' diet on the tundra is mainly arthropods and some berries. Its nonbreeding diet is mainly a mix of marine worms, crustaceans, and mollusks harvested from soft mud in shallow water. Females may feed in deeper water than the males. Breeding birds travel to coastal wetlands and flats to feed during the day.

Behavior: The male of this monogamous species attracts a female by conducting a noisy and acrobatic display flight over the undulating tundra nesting territory. Nonbreeding birds roost in flocks, often with other large shorebirds.

Nesting: The nest is a depression in the tundra, lined with plant materials, hidden by grass, and situated on a slight rise. The female lays four olive eggs marked with some darker spotting. After the eggs hatch, both parents look after the hatchlings, leading the fledglings to a marshy foraging site where they will stay until the young are able to fly.

Conservation: The IUCN Red List classes the species as Near Threatened. The population that departs Alaska in the autumn is estimated to approach 100,000.

Mid-twentieth-century ornithologists recognized the trans-Pacific movements of the Bar-tailed Godwit but assumed the birds made multiple rest stops in transit. It took twenty-first-century satellite technology to prove the fantastic nonstop journey from Alaska to southern Australasia (New Zealand, Tasmania, and eastern Australia).

Bar-tailed Godwits wintering in Australasia have been documented reaching ages of more than 23 years. Such birds have successfully made the trans-Pacific crossing at least 40 times! New Zealand–wintering Bar-tails, on average, will have traveled as much as 300,000 miles in their lifetime, and long-lived individuals more than 374,000 miles.

And, drum-roll please, here's the record-breaking overwater migratory achievement (as reported by Graham Readfearn in the *Guardian*): in 2022, Bar-tailed Godwit #234684 departed from Alaska on October 13, flying nonstop for 265 hours (11 days), traversing 8,435 miles before touching down in Ansons Bay in northeastern Tasmania. A 5G satellite tag had been affixed to its back by the Max Planck Institute for Ornithology's bird tracking project. The data demonstrated that this bird and its flock were juveniles, making the first flight of their lives. The flock passed west of Hawaii and then over Kiribati on October 19. The tagged bird passed over Vanuatu and then east of Sydney, Australia. On October 23, the bird and its flock took a sharp turn right and headed west to arrive in Tasmania on October 25. These birds traveled 763 miles per day. Their flight speed averaged 32 miles per hour for this record-breaking traverse of the Pacific. To accomplish this, these birds carry the heaviest fat loads of any shorebird, more than double their premigratory weight.

Most reference books cite the Arctic Tern as the greatest of the super-migrators. Indeed, the Arctic Tern may carry out the longest annual movement of any bird (estimated to be no less than 57,000 miles over the calendar year). But the tern does not make long nonstop overwater fasting flights; it repeatedly stops, rests, and forages on the open ocean during its seasonal migrations. Also, the tern's huge mileage count includes its considerable east–west and west–east wintering movements. Unlike the land-based godwits, Arctic Terns are, like phalaropes, essentially pelagic foragers, remaining out on the open ocean to feed and loaf for months at a time, being able to sleep each night while resting on the surface of the sea. We can salute the Arctic Tern for this achievement, but it does not begin to compare with the Olympian overwater nonstop flights carried out by the Bar-tailed Godwit.

How did such a long-distance overwater migration of the Bar-tailed Godwit evolve? Evolutionist Ernst Mayr suggested that during periods of lowered sea levels, the Pacific featured more island "stepping stones" that early shorebird migrants depended on. With rising sea levels, natural selection "proved" that this long-distance migration was, indeed, sustainable because of the substantial benefits of the rich wintering conditions in the Southern Hemisphere's summer. The two great opportunities for building fat stores and breast muscle occur in the late northern summer of the far north and the late southern summer of the far south, with long days and hyperproductive mudflats for foraging. The epic migrations of these birds continue to astound us, even as we learn more about how they do it. ❋

PART II
OUT ON THE ROAD

An adult female Long-billed Curlew in her high plains nesting habitat.

3 Solo Travels

I can count on the fingers of one hand the red-letter days when I have been privileged to see this rare and handsome wader [the Hudsonian Godwit]. *It has always been among the great desiderata of bird collectors. Its eggs are exceedingly rare in collections. Many ornithologists have never seen it in life.*

—Arthur Cleveland Bent, *Life Histories of North American Shore Birds*

THE PLAN FOR MY FIRST FIELD TRIP was to track Hudsonian Godwits migrating northward in spring across the midsection of the United States and Canada. Along the way, the godwits would lead me to wetlands and other habitats that hosted an abundance of shorebirds of many species during spring staging and stopovers mid-continent. I departed Bethesda, Maryland, on April 20, 2019, in my Nissan Xterra, heading to the Mexican border. Day one (835 miles) to Eutaw, Alabama. Day two (623 miles) to High Island, Texas. Day three (416 miles) to Arroyo City, Texas. These three long days of driving put me in the Lower Rio Grande Valley of Texas to meet the northward journeying Hudsonian Godwits arriving in the United States of America from their South American winter home.

Highlights of my 1,874-mile sprint to the southern border included some excellent BBQ at the Shed, outside of Ocean Springs, Mississippi, where I dined with ornithologist Mary LeCroy; some spicy boudin sausage picked up at Billy's, near Scott, Louisiana; and my first dose of shorebirds at Fort Travis Park at the southern end of the Bolivar Peninsula in East Texas.

The huge brick fort, named for William B. Travis (one of the hallowed defenders of the Alamo), was established by the Republic of Texas in 1836 to guard the entrance to Galveston harbor. When I arrived at the height of spring, the vast expanses of mowed lawn inside the brick fort hosted numbers of confiding shorebirds—loafing, sleeping, or foraging. Wet spots on the lawn attracted flocks of foraging dowitchers (mainly the rusty-breasted Long-billed). Scattered about were other, more enticing species. Here I saw singleton Long-billed Curlews and Whimbrels, two of the Magnificent Seven! The Whimbrel was quite shy and nervous on the ground, retreating short distances by wing when approached for photographs. I managed to sidle up close to a big female Long-billed Curlew hunting for hidden prey in the short grass. Her gracefully decurved bill was seriously long—scimitar-like. There were also Ruddy Turnstones and some unidentified peeps, the smallest of the calidridine sandpipers, near the seawall, keeping to themselves. And, of course, black-hooded, red-billed, Laughing Gulls loafed about; this is a favorite site for them because of its proximity to their nesting islands in the North Galveston boat passage. Orange-billed, breeding-plumaged Royal Terns flew overhead along the sea wall. Commonplace and noisy Great-tailed Grackles were everywhere. That big grackle dominates the human-built landscape in East Texas.

In spring, Fort Travis is one of the best places in North America to look for shorebirds. The entire green space lies within a network of narrow concrete driveways, so the expanses of lawn can be scanned from inside the car, avoiding problems of backlighting or rain. Also, for some unknown reason, the shorebirds appear more

Marbled Godwit *Limosa fedoa*
Other names: Straight-billed Curlew, Brant-bird, Badger-bird, Marlin, Brown Marlin, Big Marlin

The Marbled Godwit, our largest of the genus, is one of the Magnificent Seven.

Appearance: This godwit's plumage, which changes little between seasons, resembles that of a Long-billed Curlew—the vermiculated mantle, the rufescent underparts, and the cinnamon underwings, in particular. It is an oversized, long-legged, and long-billed shorebird.

Range: This godwit mainly summers on the northern Great Plains, and it also has tiny breeding populations at James Bay and on the Alaska Peninsula. It winters on the Atlantic, Pacific, and Gulf coasts and southward to the coasts of northern South America.

Habitat: The Marbled Godwit can be found with the Long-billed Curlew both on the breeding grounds and wintering grounds. They are most often seen in winter flocks on coastal flats. The Marbled Godwit breeds in sparse grasslands that lie near ponds and wet grasslands, as well as on tundra in Alaska.

Diet: Its diet consists of various arthropods, including crustaceans, and mollusks, which it harvests by carefully picking at the water's surface or rapidly probing mud for buried prey; while on autumn migration, it feeds extensively on aquatic plant tubers.

Behavior: The monogamous male carries out a vocal display flight over its nesting territory. The species forms large flocks at coastal wintering sites.

Nesting: The nest, hidden in short grass, is a shallow depression lined with dry grass and set near water. The female lays four clay-colored eggs, blotched and scrawled with brown.

Conservation: The Marbled Godwit is on the NABCI Watch List due to loss of prime nesting habitat and past hunting impacts. The species' population is approximately 170,000.

confiding in this environment, full of picnickers and playing children as well as birders.

Over the years, there has been no place on Earth where I have spent more quality time, at close range, with members of the Magnificent Seven. On this day, three of these "big seven"—Whimbrel, Long-billed Curlew, and Marbled Godwit—worked the grassy expanses inside the fort. Several pairs of the godwits, glistening buff and rusty, moved slowly over the lawn. These big shorebirds—the largest of the godwits—love to forage in the short grass in pairs, slowly pacing, poking, and probing for invertebrate prey. They moved with a regal air, and typically allowed me to spend time in close association as they fed. Photographed down at eye level, the godwit pairs stood out handsomely against the framing of the rich green of the lawn. These birds were in their breeding plumage, heavily marked with fine dark barring on the flanks and abundant dark spotting on the back and wings—a mix of buff and black. Their faces were marked with a dark eye-line and the crown sported a darkish cap. The long bill, with the slight upward curve typical of the godwits, was pink with a black tip. The dark legs were slim and long. This is one of our most handsome and impressive shorebirds.

After visiting Fort Travis, I raced south and west to position myself at the Mexican border to await the arrival of Hudsonian Godwits and other northbound shorebirds. On April 25 I found myself on the verge of the Lower Rio Grande Valley (LRGV) at Adolph Thomae Jr. County Park, just a bit north of Laguna Atascosa National Wildlife Refuge, in the heart of Texas thorn-scrub country, a few miles west of the Gulf of Mexico, and a few miles north of the border wall. My campsite, set near a canal, featured no prime shorebird habitat, but it was one of the few tent campsites available in southeastern Texas.

The Common Pauraque, a widespread South Texas nightjar (a whip-poor-will-like nocturnal bird), repeated its burry *pooreeyur* note through the night. The arrival of the predawn gloaming incited chorusing by a pack of Coyotes, which was followed by the low hooting of a nearby Great Horned Owl. Dawn did not come until after six thirty. As I slowly awakened, recalling these night sounds, it became clear that I was no longer in Bethesda, Maryland. Happily, the little campground was quite birdy at dawn, with singing Cactus Wrens, displaying Golden-fronted Woodpeckers, and a Long-billed Thrasher scuffling in the dry leaves.

That morning, I drove south, wending my way through the eastern verge of sprawling Brownsville and then taking Texas State Highway 4 east to Boca Chica, where the Texas-Mexico border meets the Gulf of Mexico. Much of Highway 4 passes through a variety of wetlands and salt pans—great shorebird habitat. But it was late morning, and the sun was frying the open landscape, making observing difficult. Still, some effort produced a decent list: 1 Whimbrel, 4 Ruddy Turnstones, 2 Stilt Sandpipers, 5 Sanderlings, 5 Dunlins, 1 Baird's Sandpiper, 8 Semipalmated Sandpipers, 15 Short-billed Dowitchers, 5 Willets, and hundreds of unidentified peeps at long distance. The highlight of this day was a female Peregrine Falcon stooping on a mixed flock of shorebirds, which detonated off the flats, retreating in every direction. The falcon emerged claws empty. The handsome falcon was a treat to see, but where were the Hudsonian Godwits?

Problem was, I had only once in my life glimpsed Hudsonian Godwits in spring migration, so I did not know the specifics of their habitat preferences. Yes, eBird shows where these godwits have been reported, but that is typically a coarse measure of habitat specificity. I had to use a process of trial and error. At this point, in late April 2019, as it was early in the season, not many Hudsonians were being reported to eBird in the Lower Rio Grande. And I had no close encounters with other birders, who could provide intelligence on this rare species. Another problem was habitat. The vast planar region at the bottom of Texas is heavily developed for row-crop agriculture and cattle grazing, as well as industry. Prime shorebird habitat is relatively rare. Monocultures of farm plantings share the landscape with scattered arrays of giant wind towers. This was the challenge on the sixth day of my birding journey.

The next morning, my birding destination was South Padre Island, recommended to me by the head of the American Birding Association, an expert on the LRGV. Padre Island initially appeared to be an unlikely birding hotspot. The long barrier island, much like Ocean City, Maryland, or Atlantic City, New Jersey, was loaded chockablock with beach-honkytonk development. It took some time to find out where the birds were: (1) the World Birding Center, (2) the green space behind the convention center, and (3) the flats just north of those two excellent birding spots. At the flats there were waterbirds of all types—terns, gulls, and shorebirds. Just like at Ocean City, the patches of good birding habitat lie adjacent to the horrid urban crush, and it's just a matter of locating the good spots. Here, there were most of the shorebird species I had seen the day before, plus a pair of Upland Sandpipers ("Uppies") that circled the flats north of the convention center. A group of three Baird's Sandpipers was another notable find. But no godwits this day. Still, with the addition of the Uppies, I had now logged four of the Magnificent Seven—a step forward.

Baird's Sandpipers, named for Spencer Fullerton Baird, an ornithologist who served as the second secretary of the Smithsonian Institution, are among the most sought after of the smaller shorebirds. To an East Coaster, any day you get Baird's on your list, you are having a good day. Baird's is one of the larger *Calidris* sandpipers, and it is a bit of a challenge to identify; it is often uncommon or absent from shorebird flocks. It is most like the more commonplace White-rumped Sandpiper but buffier, long-winged, and generally found in drier or grassier verges of wetlands. Baird's Sandpipers

Baird's Sandpiper *Calidris bairdii*
Other names: Bull-peep, Grass-bird

Baird's Sandpiper is a large and long-winged peep that to East Coast birders is ever-elusive and difficult to identify. Most northbound spring migrants travel up through the Mississippi drainage. The best season to see the bird on the East Coast is autumn.

Appearance: Baird's features black legs, a short, straight black bill, and long primary projections (longer than the tail when perched). It shows buffy coloration on the throat and cheek. Juvenile birds exhibit a very scaly back.

Range: Baird's Sandpiper summers in Arctic Canada, Alaska, eastern Siberia, and northern Greenland, wintering in the Andes and southern South America. In migration, Baird's prefers to travel through the middle of the continent, west to the Rockies.

Habitat: The species breeds on dry coastal and upland tundra. Migrants visit grassy shallows, mudflat edges, and short-grass.

Diet: The diet of the species is mainly arthropods on the breeding ground. In migration it takes arthropods and other small invertebrates. The species forages by rapid pecking.

Behavior: The male makes a vocal display flight high over the breeding territory, with fluttering wings and a trilled call. Migrants typically keep apart from other species of foraging peeps. Identification is aided by the bird's distinctive voice: a low and raspy *kreep!*, which is less rich than the call of the Pectoral Sandpiper.

Nesting: The nest is a depression on the tundra lined with a variety of plant matter and, in some instances, is hidden under a clump of grass. The female produces four eggs in as many days. The four pink-buff eggs are blotched with darker colors. Both adults incubate, and both tend to the young after hatching, but the female departs before the male.

Conservation: The IUCN Red List classes the species as Least Concern. The global population is estimated to be 300,000.

are generally found as singletons or in small single-species parties, usually apart from the flocks with multiple peep species.

The following morning, I rode a circuit of the wildlife loop at Laguna Atascosa National Wildlife Refuge by bicycle and spotted a single Whimbrel, some dowitchers and yellowlegs, and a big gobbler Wild Turkey, but not much else. This refuge is not famous for its shorebirds, but instead it is where a remnant US population of Ocelots hangs on by their sharp toenails. A refuge field staffer who was carrying a portable Yagi radio antenna confirmed she was checking on the location of a radio-collared cat that was hidden in the nearby thorn scrub. She assured me she had never seen this cat during the day as it is a creature of the night. She also told me some 60 Ocelots remain in Texas. Most publicity about this charismatic spotted wildcat highlights that it is an "endangered" species. But that applies only to the population in Texas. The range of this tropical cat only barely creeps northward into the United States. In fact, the Ocelot is one of the most wide-ranging cats of the Americas, occurring south to Argentina. The IUCN Red List assesses this widespread species as Least Concern—its lowest threat ranking.

My final day in the LRGV included another visit to Boca Chica and adjacent South Padre Island. Along Highway 4, I made contact with a local rarity, Botteri's Sparrow, and came upon a historical road sign, which notes the site of the last military engagement of the Civil War—the Battle of Palmito Ranch, which took place 12 days after Lee's surrender at Appomattox. Here a Union private from an Indiana infantry regiment was the last soldier killed in this bloody war.

Back out at Boca Chica there were the otherworldly SpaceX installations, where Elon Musk was planning rocket launches not far from the beach. What looked like a 1950s spacecraft, clad in shiny aluminum and standing on three legs, was perched on what may have been a launch site. I saw no sign of a real rocket on that day. Returning to South Padre Island, I discovered that the back portico of the convention center was filled with birders watching groups of migrant birds in this small patch of trees and greenery. The preceding night had produced a big trans-Gulf arrival of songbirds. The little copse of trees was vibrating with colorful birds—tanagers, grosbeaks, warblers, and vireos. Six Indigo Buntings foraged together on the ground under the shade of a bush. A female Cerulean Warbler drank at a muddy wet spot at the feet of birders under the covered portico, eliciting considerable excitement. Moments later, a Townsend's Warbler (a colorful western species) at eye level in a low tree created a new wave of pointing and excited murmuring.

The nearby flats produced oystercatchers and Pectoral and Stilt Sandpipers. A single Long-billed Curlew passed overhead—any encounter with this huge rusty-colored shorebird is a special event. As I looked out over the bay, counting shorebirds, two adult male Scarlet Tanagers caught my attention as they raced northward at eye level toward the convention center. These were songbirds that had just spent 18 hours crossing the Gulf of Mexico. They were now looking for a place to crash, get some water, and perhaps something to eat. This was migration in the raw.

Moving about in the greater Brownsville environs on this trip, I quite regularly encountered the new and imposing brown bollard wall that marks our border with Mexico. It is very tall and vertically louvered. One can poke a hand through the gap that separates each adjacent pair of vertical slats that constitute the barrier. The fence zigs and zags all around, not following the border itself because of the need to accommodate buildings, canals, and roadways. For instance, to visit the Sabal Palm Audubon Sanctuary, one drives south through a gate in the border fence even though the reserve lies within the United States proper. Construction of a large border wall in an urban area involves lots of conflict and compromise by the federal government and private landowners.

On this afternoon, eBird reported a fresh sighting of an adult Hudsonian Godwit a few hours north of my campground. Six nights

An adult breeding Least Sandpiper in its favored muddy foraging habitat.

in the LRGV produced zero Hudsonian Godwits. I packed up my gear and initiated my northward drive toward Canada and Godwit breeding habitat. It was time to find one of these elusive migratory birds.

The eBird record led me to a saltpan just north of the town of Tradewinds and a bit southwest of Bayside, not too far from the Gulf coast on the north side of Corpus Christi Bay. There were shorebirds scattered all about the flats below me. The ferocious afternoon sun was beating down on the white saltpan. I started to count the birds: 75 dowitchers, 30 Dunlins, 15 Pectoral Sandpipers, 15 Ruddy Turnstones, 15 Wilson Phalaropes, 9 Greater Yellowlegs, 9 Sanderlings, 6 Least Sandpipers, 6 Willets, 5 Lesser Yellowlegs, 5 Stilt Sandpipers, 5 Semipalmated Sandpipers, and 4 Baird's Sandpipers. And several singletons: Whimbrel, Long-billed Curlew, and Spotted Sandpiper. It was a fine list. But wait, there was a big, dark shorebird high overhead, coming toward me. Long pointed wings, long bill, dark breast, and black underwings—a spring male Hudsonian Godwit! I danced a victory jig beside my car. I had properly started on my godwit journey. And that was my fifth member of the Magnificent Seven for the trip.

Back at the saltpan the following morning, there were lots of shorebirds but no Hudsonian. The place was alive with birds: Wilson's Plovers, Snowy Plovers, Black Terns, White Ibis, and Whimbrels. And on the drive to the shorebird spot, raptors appeared in the sky: Crested Caracaras, White-tailed Hawks, and Harris's Hawks. Don't let anyone tell you that South Texas is not birdy in spring. But my very first Hudsonian Godwit must have slipped away in the night.

After stopping in Austin and then San Antonio to give bird talks to local Audubon societies, I continued my journey northward. I crossed the Pedernales River in Johnson City, where Lyndon Johnson grew up, and then made a brief stop at Balcones Canyonlands National Wildlife Refuge, west of Austin, where I photographed a Black-headed Vireo, a local endemic of Texas hill country. Another highlight of the stop was a gorgeously colored Texas Coral Snake, small, confiding, and docile—but with a deadly bite. The Scarlet Kingsnake looks much like the coral snake but is nonvenomous. Since it can be difficult to recall which color pattern belongs to which species, it is best to leave any small, brightly banded snake alone.

Central Oklahoma was rain-drenched—flooded fields in every direction, with some back roads blocked by standing water. After a night in a decrepit motel in Enid, Oklahoma, my destination was Great Salt Plains State Park in the town of Jet. East of Jet, on May 7, there was a wet and muddy plowed field filled with a big flock of pink-tinged Franklin's Gulls—migrants on their way to their breeding grounds northwest of here in the prairie potholes (a glaciated section of the northwestern United States and western Canada that features an abundance of small kettle ponds). There were largish, dark birds mixed in with the gulls. Among the more than 200 Franklin's Gulls was a mix of shorebirds: 4 Whimbrels, 2 Upland Sandpipers, 6 Stilt Sandpipers, 25 dowitchers, 15 Lesser Yellowlegs, more than 100 Wilson's Phalaropes, and—wait for it—27 Hudsonian Godwits! Their heads were down, and their bills were drilling into the inundated mud of the field. The long lens on my camera sucked up scores of images of the godwit flock as the birds fed in the shallow water.

This spot near Enid, Oklahoma, provided the key to locating spring migrant Hudsonian Godwits midcontinent—wet bare-earth plowed fields, especially ones with shallow pools of standing water. Moreover, it appeared that Franklin's Gulls often settle in fields that are favored by these godwits. Since the pink-breasted gulls were visible from a great distance, they were a great marker for fields that might feature foraging godwits.

As I prepared for my big trip north along the 100th meridian, my biggest question had been, where I would find the migrating godwits? Would they be confined to the refuges and other protected areas? Nope, not in the refuges, but mainly in flooded agricultural fields on private land. This was good news, of course, because there are a lot more privately owned flooded fields than refuge wetlands across the middle of the country.

The Hudsonians were foraging mainly in flocks, moving about together, but they would also break into subflocks. Most of the birds were probing the water to get their long bills into the mud below; undoubtedly, they were probing sewing-machine-like to feel with their bill tips for invertebrates. The birds were distant, so it was not possible to see what they were consuming. Out in the fields, these birds stood out because they were dark and large and long legged and long billed. They were easy to pick out within a mixed flock of waterbirds, and that was good news.

Foraging

Four types of foraging have been documented in scolopacid shorebirds. The first is pecking at surfaces, exhibited by Upland

and Spotted Sandpipers, Willets, the shanks (a group of slim and long-legged sandpipers, sometimes called "tattlers"), and Baird's Sandpiper. The second is the probing of soft substrates, exhibited by snipes, woodcocks, and many typical sandpipers and godwits. The third involves running with the bill in the water to capture active aquatic prey, a technique employed by some of the shanks. The fourth, rapidly pecking at the water to capture tiny invertebrate prey, is practiced by the phalaropes and some *Calidris* sandpipers.

Many sandpipers forage while standing in water—short-legged forms in shallower water, long-legged species in both shallow and deep water. Dowitchers are famous for standing belly-deep in water and thrusting their long bills into the muddy bottom for hidden prey, their heads often disappearing from view. Godwits do this too. By contrast, the Baird's Sandpiper tends to stay out of the water and pick

An American Woodcock adult in prime habitat on forest floor.

prey from drying ground or short-grass flats near water. The Buff-breasted Sandpiper is infrequently associated with water, instead favoring barren and sparsely grassed paddocks and fields, or even plowed agricultural fields. In southern Louisiana and eastern Texas in spring, flooded rice fields attract long-legged shorebirds—especially Whimbrels, Hudsonian Godwits, dowitchers, and yellowlegs. Godwits, among others, can use rhynchokinesis, the forceps-like flexible opening and closing of the upper and lower bill tips to capture submerged prey.

Researchers have found that shorebirds with specialized recurved and decurved bills require additional supporting skeletal structure. This reduces available space for a large and mobile tongue. As a result, the curlews and godwits cannot simply pull down the food item held at the bill tip; they must instead throw the item down into the throat by rapid jerking head movements. By contrast, the straight-billed snipe and woodcock can more easily draw the food item down the gullet thanks to their larger and longer mobile tongue. Moreover, a snipe or woodcock can ingest prey without having to remove its bill from the substrate.

The eyes of snipes and woodcocks sit high on the head; this prevents the birds from being able to see the tip of their bill. Although this eye position greatly aids vision above and outward, which would help with predator detection, these birds must depend upon tactile rather than visual cues to pinpoint prey.

Another foraging adaptation relates to eyesight. Sandpipers have smaller eyes than plovers do. This difference may be because plovers depend more on vision to detect surface prey, while sandpipers use their tactile bill sensors to detect prey hidden in sand or mud. Note also that the retinas of sandpipers have a lower density of rods and a higher density of cones than plovers. This gives sandpipers less visual acuity in low-light situations but provides better vision in full daylight. Sandpipers also have better color discrimination. But how that last factor aids sandpipers is not known. Most remarkably, some *Calidris* sandpipers seem to be able to use their bills to sense the presence of buried worms and mollusks that lie close by. The details of this sensory mechanism are not understood.

The feeding schedules of coastal-dwelling shorebirds depend on the tides. During ebb tide the shorebirds move out to foraging flats for their daily feeding. During high tide, these birds retreat in flocks to islands and other high areas where they can sleep, preen, bathe, and socialize. In the Far North in summer, when day length is very long, each daily cycle provides two low tides and substantial foraging opportunities. The long Arctic days encourage the mass reproduction of marine and aquatic invertebrate prey as well as long hours of productive shorebird foraging.

I set my tent in a level spot below a bluff with some protective vegetation in Great Salt Plains State Park, near the Salt Fork of the Arkansas River. In the afternoon I visited the Sandpiper Trail of the nearby Salt Plains National Wildlife Refuge. There I found hundreds of Wilson's Phalaropes twirling in the shallow water, along with foraging American Avocets and Black-necked Stilts. The huge flock of phalaropes was a nice surprise. All of them were bobbing in shallow water, showing their strongly aquatic predilections. The avocets towered over the much smaller phalaropes. Traveling a bit further north, I came upon some nice short-grass habitat hosting four Whimbrels and a pair of Upland Sandpipers. When taking a break from northward migration, shorebirds drop into all manner of habitats, based on their apparent promise of productive feeding and access to safe loafing and resting sites.

The park superintendent dropped by my campsite in the afternoon to warn me of an approaching supercell that would be overhead that evening. I got to bed in my tent early, falling asleep listening to the night sounds—Great Horned Owls, Barred Owls, Chuck-will's-widows, and Coyotes. I was awakened at midnight by the storm. My tent was buffeted by heavy rain and wind, and I was kept awake by the flashes of lightning and loud thunderclaps, but my tent kept me dry. The next day, the superintendent told me the storm had dumped six inches of rain in places, spawned a tornado nearby, and produced 80-mile-an-hour winds. It wasn't surprising that there was not much sleep for me the preceding night.

The storm left lots of standing water in the fields around the park. I rose early and drove through the countryside. The first surprise was a big American Badger loping across the road on its short legs. Then I came upon a roadside Coyote, and then good numbers of foraging shorebirds visible from the car. Perhaps these birds had been knocked out of the sky by the night's wild weather: two small parties of Hudsonian Godwits, one Marbled Godwit, an Upland Sandpiper, and an additional six species. The badger, a mammal new to me, was larger than I expected and strangely shaped, long of body and dark looking as it humped across the rural road. This was a mammal that had long eluded me. At the end of the day, a return to the flooded godwit field near Enid produced 8 Hudsonians, 25 dowitchers, 2 American Golden-Plovers, and 250 Franklin's Gulls.

The back roads of rural Alfalfa County, Oklahoma, are filled with wide-open spaces, vast fields, and verdant pastures. Of course, most of the landscape is human managed for production: cattle grazing, oil extraction, and agriculture. The low rolling hills are handsome, but the little agricultural settlements give the distinct impression of rural poverty—so many old farm buildings in disrepair. The nearest hamlet to my campground, Nescatunga, seemed to be hanging on for dear life. The main feature dominating the region is Great Salt Plains Lake, a reservoir formed by the damming of the Salt Fork of the Arkansas in the 1930s. An island in the lake supports a breeding rookery of herons and other waterbirds. I was in the heart of the Mississippi Americas Flyway at the height of spring.

Shorebird Flyways

The International Wader Study Group has identified eight global flyways for shorebirds around the world ("wader" is the British term for a shorebird). A flyway is a migration route followed by large numbers of birds on an annual basis. Recognizing flyways is important because it allows governments and nongovernmental institutions to focus resources on these pathways to conserve sites important to migrant birds. Preserving staging and stopover points along these flyways provides a big bang for the conservation buck. In addition, the flyways indicate intercontinental linkages between nations and hemispheres, thus promoting international conservation cooperation. And flyways help birders know where to spend their birding weekends in spring and fall.

Three of these shorebird flyways are in North America. The Pacific Americas Flyway links Alaska with the Pacific coast of North America, Central America, and South America. The Alaskan-breeding population of Whimbrels follows this flyway when traveling from western Alaska, to Chiloé Island, Chile. The Mississippi Americas Flyway links Arctic western Canada with the Great Plains, the Gulf of Mexico, and southern South America. The Buff-breasted Sandpiper uses this flyway on its travels from the Pampas of Argentina north to its breeding ground in Arctic Canada. Finally, the Atlantic Americas Flyway links Arctic eastern Canada and Hudson Bay with the Atlantic coast of North America, Central America, and South America. The *rufa* subspecies of the Red Knot uses this flyway when traveling from its wintering ground in southern Argentina north to Arctic Canada.

The five other global shorebird flyways are also worthy of note: the East Atlantic Flyway ranges from Arctic central and eastern Canada and Greenland to the coastlines of western Europe and the Atlantic shores of Africa to the western shores of South Africa. This is a route taken by some breeding populations of the Sanderling. The Black Sea/Mediterranean Flyway ranges from western Siberia and eastern Scandinavia south through the Black Sea and Mediterranean

Sea to central Africa. This is a migration route used by some populations of the Ruff. The West Asia/East Africa Flyway ranges from central Siberia south to western Asia, Arabia, and eastern and southern Africa. Some populations of the Ruddy Turnstone utilize this flyway.

The Central Asia Flyway ranges from east-central Siberia south through central Asia to India. Some populations of the Eurasian Curlew travel this route. Finally, the East Asia/Australasia Flyway ranges from Alaska and eastern Siberia south through East Asia and Island Southeast Asia to New Guinea, Australia, and New Zealand. This is the route made famous by the Bar-tailed Godwit, whose epic travels were discussed in the preceding chapter. The complex and divergent migratory movements of the Hudsonian Godwit indicate how fluid the flyway concept is.

After a detour to Bartlesville, Oklahoma, to tour a breeding facility for the endangered Attwater's Prairie-Chickens and to give a talk to a local Audubon bird club, I continued my trip northward to track migrating godwits. From Bartlesville, my back-roads route passed through the Joseph H. Williams Tallgrass Prairie Preserve north of Pawhuska, managed by the Nature Conservancy. Small herds of American Bison were scattered across the vast undulating open landscape. American Kingbirds and Upland Sandpipers were the featured birds here. As I crossed the border into southern Kansas, my travels were interrupted by extensive flooding of the Arkansas River. Wending my way northward in the afternoon, I encountered a flooded stubble field right by the roadside, with a ramshackle house just behind the field. I could see two men working on a car in the driveway. The field was occupied by 20 dowitchers, 2 Franklin's Gulls, 2 Blue-winged Teal, and 4 Hudsonian Godwits. The godwits were in the front yard of the house, but the men at work took no notice of these special international visitors. At around six o'clock in the evening, just east of Ellinwood, Kansas, I saw another flooded field featuring godwits, 22 this time. The big shorebirds were bracketed by Franklin's Gulls. Again, the gulls were the charm!

Having run out of daylight, I stopped in Ellinwood and took a room in the recently refurbished Wolf Hotel, an ancient piece of Ellinwood architecture that was being brought back to its 1894 glory. I was the only lodger this night and found my bedroom and sitting room filled with antiques and early twentieth-century bric-a-brac. The hotel restoration was clearly seeking historical accuracy. The hallway included staged 1920s photographs of proud hunters with their scores of dead waterfowl—their day's sporting haul (but no godwits).

An early morning visit to the flooded field I visited the previous evening produced 30 Hudsonians foraging and loafing in the morning gloom. To celebrate my encounter with these birds, I made off to Granny's Kitchen on the eastern side of Great Bend, Kansas, for a breakfast featuring corned-beef hash. A midmorning breakfast is a favorite treat of mine after an early morning outing, and Granny's Kitchen got four stars. The morning's next destination was Cheyenne Bottoms Wildlife Area, just southeast of Hoisington, Kansas.

I spent the next two nights based out of a tiny no-fee campsite on the western verge of Cheyenne Bottoms. The campsite was shaded by a line of cottonwoods and some verdant thickets in this otherwise wide-open country. It's always best when one's tent is protected from direct sunlight and strong winds that prevail in the middle of the country. Being comfortable on the road depends on being careful about establishing one's campsite. Typically, I set up my tent and my folding dining table under a large rain tarp, which protected me from rain as well as harsh direct sunlight. At issue was the wind, which would threaten to tear away the large tarp if the tarp was not set up to account for the breeze's prevailing direction.

Setting up my tent, I flushed a flock of snazzy spring-plumaged Harris's Sparrows. The area was also inundated with flocks of

Chipping Sparrows and small parties of Eastern Kingbirds on the move. Songbird spring migration was in full swing in the Bottoms.

The natural sink that forms Cheyenne Bottoms is reputed to be the largest wetlands in the interior United States. It is now heavily managed for waterfowl and shorebirds, which visit by the hundreds of thousands in spring and fall. My two-day stay here was disappointing for shorebirds because the water levels after the recent storms were high (high water means limited foraging and roosting habitat—bad for migrant shorebirds). The reserve produced 11 species of shorebirds but no godwits. In the afternoon, I returned to that field in Ellinwood to get my dose of Hudsonians. I was beginning to think that Hudsonian Godwits disdain formal protected areas in favor of wet fields that were privately owned. (I later found that was not always so—on April 22, 2022, my Smithsonian colleague Gary Graves visited Cheyenne Bottoms with fellow ornithologist Mark Robbins and counted a flock of 1,185 Hudsonian Godwits.)

The Bottoms mainly featured large flocks of dowitchers but not much else notable on the shorebird front. On May 12, the Ellinwood godwit field was empty, and a visit to nearby Quivira National Wildlife Refuge was disappointing because most of the flats were flooded. Even though Cheyenne Bottoms itself was a disappointment for shorebirds, the surrounding habitat was locally productive for other birds. Each morning, the primitive campsite in the tiny woodlot was filled with voices of songbirds, and in the evening, there were serenades by Great Horned Owls and Coyotes. The best sightings of this day were landbirds. The flock of northbound Harris's Sparrows emerged from the protection of the thickets when I sat quietly, their black faces and pink legs prominent. Then there were the gloriously plumaged Red-headed Woodpeckers—their great white wing patches glowing as these birds swooped gracefully between roadside telephone poles. It seemed the woodpeckers were spending lots of time hunting the dispersing ant swarms prompted by the recent rains.

At one point, passage on a small side road was blocked for more than 10 minutes by a passing freight train—a phenomenon I recalled from my childhood in the 1960s. In the evening a walk on the gravel road through open country was perfectly tranquil. A pair of courting Common Nighthawks soared overhead. One bird (presumably the male) zoomed by the other, producing a loud, growling *roaw* sound, like a big truck applying its compressor brakes. This was followed by the ethereal rising and bubbling trill of an Upland Sandpiper's evening flight song. These were satisfying natural encounters, but I was hoping to spend time with large flocks of migrant shorebirds, stopping for a rest on their way to the breeding zone.

Spring Stopovers and Staging

Northbound shorebirds break their migratory travels to reprovision and rest at a few locations each spring. These usually include both brief adventitious stopovers and one or more staging stops, which are longer and more important to a bird's breeding success. Stopovers tend to be impromptu, triggered by inclement weather or other random processes that force the migrating birds to touch down. In most cases, stopover sites are selected on the fly and are unplanned. The bird rests, feeds, and takes in water, then heads off when conditions improve. When one finds a single shorebird in an unexpected place, this is certainly the result of an adventitious stopover.

Staging, on the other hand, involves longer stops, when groups of birds concentrate in large numbers to rest and refuel. Staging usually lasts a week or two, long enough for the birds to recover from a long, stressful flight. Staging sites are rich foraging environments, used annually, where the birds can put on significant weight to help them to arrive at their breeding grounds in top physical condition. In addition to refueling, staging involves other critical activities: physiological recovery from the stresses of a long flight, sleeping, information gathering, social interactions with future flock mates, and waiting for ideal weather for the next flight. Staging sites are great places to witness remarkable shorebird concentrations. Cheyenne Bottoms, Kansas, in spring can harbor large flocks of shorebirds, including northbound dowitchers or Hudsonian Godwits.

For the Red Knot, Ruddy Turnstone, Sanderling, and Semipalmated Sandpiper, the beaches of Delaware Bay (in Delaware and New Jersey) provide an important staging area in late May. These migrating shorebirds stop for 10 days or more to feast on the tiny eggs of Atlantic Horseshoe Crabs, which time their spawning event to the moon and tides. The Atlantic Horseshoe Crab is a major player in the lives of the Red Knot and other Atlantic migrating shorebirds. It is not a crab but rather a relative of scorpions and ticks and is known as a chelicerate arthropod (meaning it has pincers). It is one of four extant species of the family Limulidae, an ancient invertebrate lineage with fossils tracing back 440 million years. Our species ranges from Maine to the Yucatán.

The spawning of the Atlantic Horseshoe Crab is a major environmental event in the littoral ecosystems of Delaware Bay each spring. The spawning takes place in sandy shallows along beaches, with the females depositing their large payload of tiny eggs into the sand and the males fertilizing the eggs once deposited. During the height of this reproductive orgy, at a spring tide in May or early June, huge flocks of Sanderlings, Ruddy Turnstones, Semipalmated Sandpipers, and Red Knots descend on the shores of Delaware Bay, with attendant Laughing Gulls, to feed on the billions of eggs. Of course, this attracts birders in large numbers to sites such as Little Creek, Delaware, or Reeds Beach, New Jersey, to witness the profusion of life, filled with color and action. This is a classic staging site for Red Knots, tied to a predictable annual event in nature. Here the knots and other shorebirds feast with the intention of doubling their weight during their extended stay. This prepares them for the energetic demands of making their final jump to their Arctic breeding grounds and enables them to arrive in good physical condition to compete for a territory and a mate—a critical test of their fitness.

Red Knot ***Calidris canutus***

Other names: Red-breasted Sandpiper, Gray-back, Silver-back, Rosy Plover, Robin Snipe, Wahquoit, Whiting

This pudgy shorebird stages in flocks to feed on the eggs of the Atlantic Horseshoe Crab in Delaware Bay during the high spring tides in late May and early June.

Appearance: Larger and chubbier than a peep, the Red Knot, the largest North American *Calidris* sandpiper, exhibits a short but thick-based bill, short legs, long wings, and a whitish rump with fine dark barring. It is ruddy breasted in spring and quite pale in winter.

Range: The North American populations summer in the Arctic of Canada and northern Alaska, wintering on the Atlantic, Pacific, and Gulf coasts and southward as far as southern South America. Other races breed in Eurasia, wintering in Africa and Australia.

Habitat: The species breeds in barren upland tundra, often on a gravel ridge, typically near a pond or stream. Migrating and wintering birds mainly inhabit coastal and marine environments, often sharing wave-washed beachfront with flocks of Sanderlings.

Diet: The Red Knot's diet is dominated by polychaete worms, small crabs, marine mollusks, and some plant matter. It forages by pecking and lightly probing soft substrates, and it often feeds in flocks. This species will feed both during the day and night.

Behavior: The monogamous male conducts a display flight over the nesting territory, circling, quivering his wings, and singing a musical series of sad sweet notes. Nonbreeding birds are mainly silent.

Nesting: The ground nest is a shallow scrape lined with plant matter. The male prepares several nest scrapes and the female chooses one. The female lays three or four olive-gray to olive-buff eggs, dotted and blotched with cinnamon-brown. Both sexes incubate the eggs, but the male does most of this, and the female heads south first.

Conservation: Classed as Near Threatened by the IUCN. The current population in North America is estimated to be 139,000.

An adult male Hudsonian Godwit at a midcontinent migratory stopover.

In 1981, an aerial survey of Delaware Bay's beaches at the peak of the horseshoe crab extravaganza counted an estimated 1.5 million shorebirds, 150,000 of them Red Knots. Those numbers, sadly, are now history, because of overharvest of the crab. Political battles in New Jersey, Maryland, Virginia, and South Carolina to limit the commercial harvest of the horseshoe crabs to protect the Red Knot populations have raged over the years, led by the American Bird Conservancy and its partners. But the spring event still attracts lots of feeding birds and lots of birders.

The Copper River Delta in Alaska is another important spring staging site for shorebirds headed to tundra breeding grounds. Western Sandpiper is the featured species here. Other important North American staging sites include Gray's Harbor, Washington, and the Bay of Fundy, New Brunswick. Staging sites, of course, are also favored birding hotspots, where the birder gets a chance to see flocks of shorebirds in concentrations of hundreds or thousands. Birding festivals often focus on a staging site during its high season (such as the Copper River Delta Shorebird Festival in Alaska in early May each year). More importantly, the availability of staging sites is critical to the well-being of entire populations of some species, which depend on the rich fare to remain well provisioned and healthy on the way north or south. Therefore, staging areas are of major conservation significance.

On May 13 my travels continued northward 327 miles, from Kansas to the Rainwater Basin of eastern Nebraska. On the way, near Minneapolis, Kansas, cruising at 75 miles an hour, my eye was drawn to three big dark shorebirds in a roadside wetland. Jerking the car to a halt, I confirmed these were three Hudsonian Godwits in their favorite midcontinent stopover habitat—a flooded field. The godwits shared the wetlands with a single unidentified peep. One godwit pulled a long worm out of the muck then ran off to avoid the tasty tidbit being stolen by one of the other birds. Northbound Hudsonian Godwits tend not to stage in spring. Instead, they stop briefly at wetland sites like this one, staying only one or a few days before continuing northward.

I recorded details of my daily observations using a digital recorder that I spoke into many times a day, even while driving. This allowed me to get my thoughts and experiences down in full detail in real time, which prevented me from losing the freshness of the sightings with elapsed time. Every little detail went into the recorder.

My morning's drive proceeded through attractive rural country featuring some low limestone hills. A roadkill tally for the day produced an impressive list: 60 Nine-banded Armadillos, 25 Raccoons, 20 White-tailed Deer, 20 Virginia Opossums, and 2 Coyotes. Carnage!

Buff-breasted Sandpiper *Calidris subruficollis*
Other names: Buffy, Hill Grass-bird, Cherook, *Tryngites subruficollis*

This rarely seen and very comely sandpiper migrates through the middle of the continent to and from its Arctic breeding ground.

Appearance: The species is distinctive, with its unmarked buff underparts and finely vermiculated dark upperparts. The head is small and rounded with a prominent dark eye and a small black bill. In flight the bright white underwing is prominent.

Range: The species summers in the Arctic of north-central Canada and northeastern Alaska. The species winters on the Pampas of Argentina.

Habitat: The species migrates in flocks, which settle on prairies, short-grass pastures, airport grass verges, plowed fields, stubble fields, sod farms, golf courses, and dry interior flats of barrier islands. It breeds on tundra slopes near water.

Diet: Its diet is mainly arthropods, picked from the ground. This bird is a strictly visual forager.

Behavior: This is a lek-breeding species, meaning that males cluster in courtship sites to win over females. The male gives a single or double wing flash display. It also conducts paired vertical display flights above the lek. The Buff-breasted Sandpiper is typically silent, but the call note is a bubbling musical series on a single pitch.

Nesting: The nest, set on a hummock of moss in tundra, is a shallow depression lined with plant matter. The female lays three to four buffy-grayish-white eggs that are boldly spotted longitudinally with browns and purplish gray. The males do not assist in nesting.

Conservation: The population of this species was much reduced by market hunting in that unfortunate heyday. Today it is on the IUCN Red List as Near Threatened and included on the NABCI Watch List. It has declined by as much as 60 percent since 1980. The species population is estimated to be 56,000. This is among the most elusive of our North America breeding shorebirds.

The occasional living mammal appeared, crossing expanses of plowed bare-earth fields: two Raccoons, a Striped Skunk, a pair of Franklin's Ground Squirrels, and a White-tailed Deer. There's not a lot of cover for them, but they make do somehow.

I was just south of the prairie potholes. The Rainwater Basin of Nebraska features big-time industrial agriculture. In every direction there are bare-earth fields that will grow wheat, corn, soybeans, and sorghum. Luckily, this region also features lots of undeveloped wetlands that are conserved as waterfowl production areas.

I spent four days working the back roads of eastern Nebraska: Utica, Waco, Thayer, Tamora, Gresham, and Beaver Crossing—all east of the Platte River. I checked out the wet spots for godwits, and other water lovers, and the dry fields and pastures for grasspipers (grassland-loving sandpipers, which include the Upland Sandpiper, Long-billed Curlew, Buff-breasted Sandpiper, Baird's Sandpiper, and Marbled Godwit). A flooded field north of Gresham—a small green space owned by the American Wetland Trust—was quite birdy: 4 Hudsonian Godwits, 1 Marbled Godwit, 15 Stilt Sandpipers, 2 Baird's Sandpipers, 20 Least Sandpipers, 25 Wilson's Phalaropes, 25 Semipalmated Sandpipers, 1 Dunlin, 5 White-rumped Sandpipers, 2 Spotted Sandpipers, and 4 Lesser Yellowlegs. An adjacent field of dry corn stubble hosted three Upland Sandpipers and two flocks of Buff-breasted Sandpipers (or "Buffies"). The Buffies hid fiendishly well in the expanses of corn stubble. I had to look for subtle movement to pick them out. One male did his "raised wing" display, flashing his bright white underwing. Then another bird did his display with both wings spread wide and breast feathers puffed out. These males were practicing the moves they would deploy when they arrived on their tundra breeding sites. When flushed, they were swift and jet-like, their white underwings flashing in the sun. Then they quickly dropped back to the ground, disappearing in the stubble to continue their foraging in virtual anonymity. It's not often that I get to spend time with Buffies on the move. This was a Nebraska treat. ❋

4 Pointed North Toward Canada

When bottom lands brighten beneath April showers, when red-wings whistle Oakalee! *Along brooks in which the alewives throng up from the sea, then the snipe blow in on the south wind by night and pitch into their green meadow world.*

—Henry Marion Hall, *A Gathering of Shorebirds*

On May 17 I drove northward 446 miles from eastern Nebraska to Aberdeen, South Dakota. The route crossed the Missouri at Yankton, South Dakota. As I sped northward on Route 81, at Marion, South Dakota, there was a mass of dark birds in a large stubble field. Swerving off the road, I parked and grabbed my binoculars. Hundreds of black-breasted birds were hunkered down nervously in the field, eyeing me. Their golden-flecked backs shone in the sunlight. Suddenly, a raptor overhead set the entire flock aflight. My photograph of this aerial flock revealed 249 American Golden-Plovers in their sublime spring breeding plumage. Even though these were plovers and not scolopacid shorebirds, this stood out as the prize encounter of the day. This plover had popped up several times on this trip, but always as single individuals, and all of them in drab olive nonbreeding plumage. These birds were stunning in their breeding black and gold.

Downtown De Smet, South Dakota, provided lunch in a friendly diner. The town features one of Laura Ingalls Wilder's later childhood homes, described in her books *The Long Winter* and *Little Town on the Prairie.* This was where schoolteacher Laura Ingalls met and married Almanzo Wilder, to whom she remained married for more than half a century. My children loved to read and reread the Wilder books and so it was a treat to spend time in Wilder country.

Further north, the South Dakota that greeted me was windy, stormy, and cold, brought on by a three-day "blow." Every patch of water was white-capped. Few trees had leafed out here. At Sand Lake Reservoir, aerial Cliff Swallows battled the cold and wind by skimming low to the water. Houghton, beside this big body of water, provided no place to camp. Facing the heavy wind, the rain, and the day's high temperature of 48°F, I retreated to the college town of Aberdeen, holing up in a motel for three rainy and windy nights.

Each morning I ventured out in search of godwits. Not far north of Aberdeen, a flooded field produced 75 Franklin's Gulls, 50 Yellow-headed Blackbirds, 3 Marbled Godwits, and 3 breeding-plumaged Hudsonian Godwits. One of the Marbled Godwits was chasing one of the Hudsonians all about, trying to snatch a prey item. Here I could see that the Marbled was a substantially bigger bird.

It was interesting to note that, to this point, the migrant Hudsonians had made no sound of any sort. Are they mute on migration? (I found on my various travels that they appear to be.) These are birds that are quite vocal on their northern breeding grounds.

On this nasty day, lots of songbirds were down in the roadside grass. It seemed like there had been a migratory stoppage, this terrible storm knocking northward-migrating birds right out of the sky. Winds are critical to migrating birds, and when large groups of northbound migrants, benefiting from southerly winds, meet up with a front of strong north winds, this creates what is called a *fallout.*

Once the storm had passed, my route took me northwest toward the North Dakota–Manitoba border. On a morning visit to Lostwood National Wildlife Refuge I encountered Upland Sandpipers and Marbled Godwits. The trip's mammal list grew. I added the Mule Deer, Red Fox, and Thirteen-lined Ground Squirrel. My circuitous route led through the petroleum-frenzy horror of the Bakken Shale along Highway 22: giant trucks, oil camps, instant towns of ready-made housing, stacked piping, gas flares lighting the sky, and the look of money being extracted straight from the ground. I longed to escape to greener pastures. Happily, at day's end I set my tent in Lake Metigoshe State Park in the Turtle Mountains.

Lake Metigoshe presented another world entirely. It was Memorial Day weekend and spring camping was just opening in this interior north country. The Turtle Mountains rose dark along the international border, cloaked with a thick dwarf forest of oaks, aspens, birches, and other broadleaf species. The Metigoshe woodlands rang with the voices of Northern Waterthrushes and Yellow Warblers. No conifers were to be seen. Still, it had a boreal feel, with the lake enlivened by displaying pairs of Red-necked Grebes rushing, side by side, across the water's glossy surface. This grebe was an unfamiliar species to me, as it is seen only as a winter vagrant along the East Coast.

Two nights at Lake Metigoshe offered a base for exploring central North Dakota. On Saturday, eBird sent me to Devils Lake to chase down reports of Hudsonian Godwits. A traverse of 256 miles produced not a single godwit of any species. It rained much of the day, limiting visibility and any joy in driving. Shorebirds on this wet and gloomy day: Wilson's Snipe, Stilt Sandpiper, Dunlin, Semipalmated Sandpiper, and Ruddy Turnstone. The shorebirding was akin to pulling teeth. My body core remained chilled the entire day.

On May 25 I crossed into Canada south of Goodlands, Manitoba. This was the middle of nowhere, and mine was the only car crossing the border. Descending out of the wooded Turtle Mountains, the

Two adult Wilson's Snipes foraging in a freshwater wetland.

view was of a vast plain with lots of open fields and a scattering of the old fashioned "nodding donkeys" pumping their crude oil. A tour of Whitewater Lake Wildlife Management Area netted six Marbled Godwits, but no Hudsonians. The flats on White Lake swarmed with nonbiting flies of some unknown variety. An assortment of shorebird species foraged, scattered widely across the mudflats, but it was impossible to get close enough to identify them.

I lunched in nearby Deloraine; the town square featured a towering metal Purple Martin nesting structure—more than 80 feet tall with more than 100 nest holes. Lots of martins were circling the structure, so the avian apartment building was serving its intended purpose. More towns should have similar nesting sites.

I spent four nights in the Kiche Manitou Campground of Spruce Woods Provincial Park, north of Glenboro, Manitoba. The park sits atop sandy glacial outwash of the Assiniboine River, which has produced ancient sand dunes—one of the park's special natural features (along with the stands of old growth White Spruce). This was hilly forested country, quite distinct from the vast planar Manitoba fields to the south toward the US border. My campsite hosted Mourning Cloak butterflies, Snowshoe Hares, and Red Squirrels, as well as Cape May and Blackpoll Warblers stopping over on their way north to their boreal breeding grounds. A male Rose-breasted Grosbeak sang relentlessly in the afternoon from atop the bluff, his deep pink breast patch glowing in the light of the setting sun.

Being on the trail of shorebirds, not grosbeaks or wood warblers, I needed to get to Glenboro Marsh, a few miles south of my campground. eBird pointed me there, promising all sorts of waterbirds, including an array of my targeted godwits. Glenboro Marsh was the focus of my shorebird-watching for the next few days. It was the breeding home of Wilson's Snipes, Willets, and Marbled Godwits, and it also featured an active lek of Sharp-tailed Grouse.

A heavy fog enveloped the marsh, and the marsh grass was stiff with frost the May morning of my first visit. Surrounded by expanses of agricultural lands, this glorious Canadian wetland at dawn was alive with birdsong, featuring the voices of Western Meadowlarks, Soras, Virginia Rails, and dozens of Marsh Wrens. Small flocks of Least Sandpipers worked the shallows, and a Spotted Sandpiper teetered at the water's edge. Individual Marbled Godwits reacted to my presence with sharp complaining notes. These birds were nesting nearby and treated me as a potential predator.

I heard a Wilson's Snipe carrying out its winnowing display high in the sky over the marsh. The bird's tail feathers produced a rapid and rising series of about 20 owl-like *hooah* notes as the bird raced across the sky. Usually heard after dark, the snipe also does its flight display in the early morning, especially when it is foggy. The male in spring also likes to perch on a fencepost and give a series of *kiiik* notes. When not producing display sounds, the snipe is usually

Wilson's Snipe ***Gallinago delicata***
Other names: Alewife-bird, Squatting Snipe, Jack Snipe, Common Snipe, *Gallinago gallinago*

Wilson's Snipe is probably the most abundant and one of the widest-ranging shorebirds on the continent. It can most often be seen in the open marshy countryside in the early morning when an individual is perched conspicuously atop a fencepost.

Appearance: Wilson's Snipe looks a bit like a cross between a woodcock and a dowitcher—it being most closely related to the latter. The snipe's head and face are boldly marked with dark stripes, as are the flanks.

Range: This bird breeds from Alaska east to Labrador and south to Arizona, Colorado, and the Appalachians. It winters across the southern half of the Lower 48. Some wintering birds end up in South America and the Caribbean.

Habitat: A rarely seen but often heard shorebird of interior bogs, wet meadows, and marshlands across North America.

Diet: Wilson's Snipe forages by drilling its long bill into the soft ground to take all manner of invertebrates, including earthworms, crustaceans, and arthropod larvae.

Behavior: Both sexes conduct an aerial display flight in which their specialized outer tail feathers vibrate rapidly when the bird dives toward the ground, producing a rapid series of quavering notes. In addition, the male will carry out a terrestrial display for the nearby female with his tail spread wide.

Nesting: The female arrives on the breeding ground first in spring. The nest is a depression on the ground hidden by a swatch of tall grass. The female lays four olive- or clay-colored eggs marked with dark brown. Both parents brood and feed the offspring.

Conservation: The Red List status is Least Concern. The species is hunted as a gamebird. The population on the continent is probably more than 2 million; 93,000 were bagged in the US in 2021.

well out of sight. Its heavily patterned plumage (dark streaked and spotted) renders it nearly invisible when it is foraging on the ground in grassy wetlands.

Early on a cold and damp day, I again toured Glenboro Marsh, this time with Lorelei Mitchell, Glenboro's long-time bird expert. I was introduced to Lorelei by the park staff when they heard I was interested in birds. Our first stop was the Sharp-tailed Grouse lek situated on a well-drained patch of nearly bare ground above the marsh. Males on the lek were posturing to each other, lowering their wings and raising their pointed tails, shifting from side to side and occasionally leaping in the air, giving a *chok chok* series of notes. Female grouse lurked around the edges, eyeing their potential mates. The displaying males inflate their purple neck sacs and erect their bright orange eyebrow combs (a fleshy outgrowth found above the eye in fowl-like birds) to impress the females. Pairs of males occasionally threw themselves at each other in bouts of aggression. By the time the sun started to peek above the horizon, display activities faltered and the birds dispersed. No copulations were observed. Lorelei Mitchell has been coming to watch these displaying grouse for decades; she never tires of this spring phenomenon.

The back roads of Glenboro were mainly dirt and gravel. Recent rains had turned many of these into muddy morasses that plastered the undercarriage and wheel wells of my car, making it nearly undrivable. Traveling these inhospitable roads produced zero Hudsonian Godwits—instead, Upland Sandpipers and Marbled Godwits—not bad alternatives, all things considered. The godwits haunt the swales of tall damp grass. The Uppies hang out in the drier and lower grass. Although the Glenboro area was mainly bare plowed fields, there was just enough unworked and marshy habitat to support populations of these species of grasspiper royalty—two of my Magnificent Seven. They eyed me warily and rarely allowed close approach. This was probably the result of generations of hunting on the breeding grounds, in migration, and in their southern wintering haunts.

One morning I made a side trip to Riding Mountain National Park, two hours north of Glenboro. The park's gateway is the lovely and antique town of Wasagaming. Set on pristine Clear Lake, which is surrounded by grand boreal conifer forest, Wasagaming is a gorgeous summer vacation destination. Lunch at the Lakehouse hotel was a delectable set of pulled-pork tacos. The Canadians know how to do things just right. Anyone wanting a bucolic holiday in the north woods would not be disappointed staying at this perfect venue.

After days of wide-open prairie lands, I now found thousands of acres of boreal spruce forest. Driving unpaved Highway 19 east to west across the park brought me happy encounters with a Boreal Chickadee, a Common Loon, and a yearling Black Bear that was hunting for fruit in the roadside shrubbery. White-throated Sparrows were singing, and a Canadian Tiger Swallowtail hunted for flowers beside Lake Katherine.

It was interesting to see how latitude and elevation influences habitat. By driving north and gaining some elevation, I saw the environment transition from prairie into boreal forest. This was a cooler zone where evaporation is reduced and the rain that falls has a greater impact on the prevailing vegetation. The park's breeding shorebirds include Wilson's Snipes, Upland Sandpipers, Spotted Sandpipers, and Solitary Sandpipers—all are lovers of boreal forest and its attendant grassy stream and wetland openings.

The habitats of Riding Mountain National Park are ideal breeding grounds for the Solitary Sandpiper. As the name implies, this bird is a loner. It loves the verges of ponds and streams in boreal forest. It is not seen as frequently as the more familiar Spotted Sandpiper, which also favors boreal forest openings. The Solitary has the look of an undersized Lesser Yellowlegs, slim and thin-billed, but with long, pale green legs. The Solitary Sandpiper is shy and flushes easily. It is a dainty forager, mainly picking at surfaces to take tiny invertebrate prey. The breeding adult is dark backed with white dorsal speckling and a fine white eye-ring. The belly and lower breast are

Solitary Sandpiper *Tringa solitaria*

Other names: Barnyard Plover, Black Snipe, Wood Tattler, *Actitis solitarius*

Appearance: The Solitary Sandpiper is a slim and dark sandpiper. The species is dark mantled, gray throated, and white bellied with a narrow white eye-ring and a short bill.

Range: This sandpiper summers in Canada and winters in Mexico, the Caribbean, and much of South America. It is a fairly common migrant across the Lower 48, but generally uncommon along the coast. Some autumn migrants depart from the New England coast out across the Atlantic nonstop to South America.

Habitat: The species nests near ponds in boreal forest. It winters at stream sides, wooded swamps, and freshwater marshes. It is very methodical in searching the verges of a small pond. It is rarely or never seen on a beach or seashore.

Diet: The species' diet is arthropods and aquatic invertebrates, harvested by picking daintily at the water's surface. When foraging, it occasionally shakes its leading foot when in water to stir up aquatic prey. It tips its tail when foraging.

Behavior: It is typically seen as a singleton foraging at the edge of a wooded pond or stream—often in places where no other shorebird except the Spotted Sandpiper might be expected. The Solitary Sandpiper is apparently monogamous. Its call is a high-pitched *tii tii tii or tititi ... tititi.* It can be seen walking while bobbing its whole body.

Nesting: This sandpiper adopts the discarded nest of a boreal songbird up in a conifer. The female lays four pale, greenish-white eggs marked at the larger end with grays and browns. Both adults incubate the eggs.

Conservation: The Solitary Sandpiper's decline since 1980 is estimated to be as much as 40 percent. The species' population is approximately 189,000.

all white, lacking the prominent black spots exhibited by the Spotted Sandpiper. The Solitary is perhaps a bit more wilderness loving than its more commonplace cousin.

Longevity

The Solitary Sandpiper I watched on the shore of Lake Katherine may have been a decade old. Shorebird mortality is high in the first year of life, but most shorebirds that survive that first year live more than five years. Small shorebirds typically live 3–10 years, though one banded Least Sandpiper was retrapped alive on its breeding grounds after 17 years, and one Red Knot was proven to be 22 years old. Larger shorebirds tend to be long lived. Most remarkably, the super-migrators appear to be among the longest surviving of shorebirds, indicating that their extreme transoceanic migratory flights do not produce high mortality among adult birds. A Whimbrel was retrapped at 24 years old. The longevity record for shorebirds overall is 32 years, an achievement accomplished by a pair of banded Eurasian Curlews.

Back in Glenboro, after a final night in Spruce Woods, I rose early and headed to the northeast to explore the shorebird hotspots around Winnipeg, following Manitoba's Pine to Prairie birding trail: Delta Beach, on the southern shore of giant Lake Manitoba; Delta Marsh Wildlife Management Area; and Oak Hammock Marsh. After a final day of birding by car in Manitoba, I would fly to Churchill, Manitoba, most famous for its Polar Bears but also a nesting ground of the Hudsonian Godwit.

Delta Beach is a popular Manitoba summer vacation community, but summer was still weeks away when I was there in late May, and the community was empty. The shoreline was mainly private, so getting access to the lakeshore was difficult. But Delta Beach

Ruddy Turnstone *Arenaria interpres*

Other names: Chicken-plover, Bishop-plover, Calico-back, Chic-aric, Creddock, Horsefoot Snipe, Sea-quail, Salt-water Partridge

The Ruddy Turnstone is one of the most widespread of the world's shorebirds.

Appearance: This distinctive-looking and compact shorebird exhibits a short neck, an abbreviated and thick-based but pointed bill, and short legs. The breeding male is harlequin patterned with black, white, and chestnut plumage and orange legs. The female is duller colored. Winter birds are quite dark and plain.

Range: The species breeds across the Arctic, from Scandinavia east to Alaska, northern Canada, and Greenland. Our birds winter on the Atlantic, Pacific, and Gulf coasts of the southern United States and southward into Central and South America. Other populations winter to Africa and throughout the coasts of Europe, Asia, and Australia. As a long-distance, transoceanic migrant, the species stays mainly near the coast in the nonbreeding seasons, but it is fairly common in migration in the Great Lakes region.

Habitat: This turnstone breeds on open ground—rocky ridges and sparsely vegetated tundra near water. It winters along coastlines of all kinds, especially where there is abundant debris to hunt in.

Diet: The diet of this species is a mix of mollusks, crustaceans, and insects, harvested by picking, probing, and turning over stones and other debris.

Behavior: Turnstones associate regularly with other shorebirds in winter. The species is monogamous.

Nesting: Both turnstone adults incubate. The ground nest is a sparsely lined shallow depression. The female lays three or four clay-colored eggs, scrawled and blotched with grayish brown.

Conservation: The North American population of Ruddy Turnstones has decreased by as much as 75 percent since 1980. The population estimate for North America is 245,000.

Campground offered a sandy beach featuring shorebirds. Lots of peeps (Baird's, Western, Least, Semipalmated, and White-rumped), Sanderlings, and Dunlins crowded the beach front.

More than 50 spring-plumaged male Ruddy Turnstones foraged as a flock on a backyard lawn at one of the summer houses on the lake. It was stunning to see these beach lovers on a neatly mowed Canadian lawn. The breeding male of this species in spring is the most brightly patterned North American shorebird. The birds, with their short legs, are low to the ground, and individuals favor turning over rocks and other items in search of hiding invertebrates. The species is most commonly seen along Atlantic, Pacific, and Gulf shores. Encountering this flock on an interior lawn was a surprise.

At Oak Hammock Marsh Wetland Discovery Centre, a walk on the boardwalk and a scoping of the mudflats produced a single American Golden-Plover, 15 Black-bellied Plovers, a single spring-plumaged Red-necked Phalarope, and 4 Marbled Godwits. Today's highlights included 12 species of shorebirds, 3 plover species, several American Avocets, and 25 Caspian Terns.

Manitoba north of Winnipeg features two gigantic lakes (Winnipeg and Manitoba), thousands of smaller water bodies, and wetlands galore. One could spend an entire spring birding the hotspots of Manitoba. But to birders south of the US-Canada border, this avian paradise is little known.

After my long day of birding, I retired to a motel in Winnipeg in anticipation of my morning flight north to Churchill. Looking back, I was pleased by the string of productive days spent in Manitoba and North Dakota, where I had seen an array of stunning landscapes populated by shorebirds, other waterbirds, upland game birds, and songbirds. Some were preparing to nest, whereas others were moving northward to their sub-Arctic breeding areas. Churchill would give me a first look at the Canadian tundra, a distinctive habitat I had long wished to experience firsthand. ❋

5 Churchill Godwits

Graceful of form and movement, wild enough to require some care to approach (only propinquity reveals the charm of these birds), they are the very embodiment of unfettered freedom, denizens of wide horizons, and almost all accomplished travelers.

—Henry Marion Hall, writing of the Scolopacidae
in *A Gathering of Shorebirds*

On May 30 I flew two and a half hours north from Winnipeg on one of Calm Air's turboprop ATR 42s to Churchill, Manitoba, the self-proclaimed "Polar Bear Capital of the World," set on the western shore of chilly Hudson Bay. The flight took me over a vast landscape of boreal spruce forest, lakes, and muskeg (Canadian bogland). I saw few signs of humankind. As the plane approached Churchill, below us lay Hudson Bay—entirely white as far as the eye could see because of the thick covering of ice that had built up over the winter. The Churchill River was also choked with ice but did exhibit some spots of open water. The sky overhead was leaden, and the landscape below was shrouded in gloom—the look of early spring in Churchill. And don't forget the snow—piled in long drifts beside the roads and in irregular strips across the landscape.

In Winnipeg the preceding afternoon, the temperature had reached 80°F. But here in Churchill, I would need extra layers!

The Churchill landscape is a mix of tundra, abundant lakes and ponds, boglands, and patches of spruce forest. This is Canada's southernmost tundra, a result of the freezer-like properties of massive Hudson Bay, which is ice covered from late October to late June. The big block of ice that is Hudson Bay chills the adjacent landscape, giving it an Arctic look and feel. That is why these shorelands are home to Polar Bears, which in most places are only found far to the north.

Churchill was my destination because it is the most accessible place where Hudsonian Godwits nest. And at that time of year, the godwits were settling in and just starting the breeding cycle. What makes Churchill an ideal place to study godwits is the presence of the Northern Studies Centre, my base of operations for the next week. The Centre provided lodging, a research library, three daily meals, a loaner vehicle, and even a shotgun-toting guide—everything I needed to get up close and personal with nesting Hudsonian Godwits on their breeding territory.

There is one practical issue facing researchers in Churchill—Polar Bears. The Centre has a general prohibition against visitors wandering the landscape because of the chance of bumping into a prowling bear. One must stay near the car or the roadside or else have an armed companion looking out for bears.

My first afternoon in Churchill I took out my ancient car—a beat-up Chevy Suburban (quite suitable to my needs). The roads were nearly all unpaved, so travel through the tundra landscape was slow and careful. The temperature was in the 30s; the sky was gray. But birds were everywhere. Pairs of Pacific Loons floated on openings in ice-studded ponds and lakes. Flocks of Snow Geese, including many blue-morph individuals (those with blue plumage known informally as "Blue Geese" as opposed to the Snow Geese with their white plumage), foraged on patches of snow-free tundra. Redpolls flitted about the spruces, singing their spring song. And shorebirds frequented the wetlands.

The road system in Churchill is simple. There is a road from the Northern Studies Centre west to downtown and the mouth of the Churchill River. Just east of town the Goose Creek Road leads south, passing by the airport and through the community of Goose Creek. It continues south along the eastern bank of the Churchill River. There are a total of 25 miles to ramble about in search of wildlife. A small additional track, the Twin Lakes Road, travels south from the Northern Studies Centre, but at the time of my visit it was impassable because of drifted snow.

The most productive wetlands for me were along the Goose Creek Road, south of town. There was a series of shallow ponds and mudflats strung out on the east side of the road, and these hosted small clots of shorebirds. Most of these were likely males, given the

Male (*left*) and female (*right*) Red-necked Phalaropes on sub-Arctic breeding habitat.

Stilt Sandpiper ***Calidris himantopus***

Other names: Bastard Yellow-legs, Stilted Sandpiper, Mongrel, Green-leg, Long-legged Sandpiper, Frost Snipe, *Micropalama himantopus*

The Stilt Sandpiper is uncommon enough to be appreciated every time it is encountered.

Appearance: This is a tall and trimly plumaged sandpiper. The chestnut trim on the cheek and crown of the spring adult makes this a very handsome bird in season. Note the slightly decurved bill. Winter birds are quite plain but exhibit a diagnostic white eyebrow. In flight, these winter birds look quite like Lesser Yellowlegs.

Range: The species summers in northeastern Alaska, Arctic Canada, and the western shore of Hudson Bay, wintering from coastal Texas and South Florida and the Caribbean to southern South America. In both spring and fall, most birds migrate through the interior of the Lower 48.

Habitat: This sandpiper breeds in sedge meadows in tundra. Migrants frequent shallow pools and shallows, often in association with yellowlegs and dowitchers. It prefers freshwater environments.

Diet: The Stilt Sandpiper's diet is arthropods, mollusks, and some plant matter. It often forages up to its belly in deep water. When foraging, the birds point their bill straight down, which is distinctive.

Behavior: This is a monogamous species. The male conducts spectacular aerial displays for the female. The species is largely silent.

Nesting: The nest, set atop a dry ridge surrounded by water, is a shallow depression in the tundra vegetation. The female lays four pale grayish-buff eggs boldly spotted with brown and purplish gray. Both sexes incubate the eggs.

Conservation: The species has declined by as much as 70 percent since 1980. A 2008 estimate of the global population was 1.2 million. At least one cause of decline appears to be the increase of Snow Geese, which have damaged the breeding habitat.

current weather conditions; only the males would brave this cold environment, arriving early to snag a good breeding territory. A scoping of the flats produced 40 Stilt Sandpipers, 15 Dunlins, 5 Ruddy Turnstones, 5 Lesser Yellowlegs, singletons of Least Sandpiper, Short-billed Dowitcher, Wilson's Snipe, and Spotted Sandpiper—and 6 Hudsonian Godwits. My target species was there despite the iced-over river, snow drifts, and cold. And indeed, the six godwits were brightly plumaged males.

The Stilt Sandpipers were also there in numbers, and they were all featuring their pretty spring plumage—the heavy dark patterning on the back, belly, neck, and flanks; the rusty patches on the cheek and side of the head; the prominent white eyebrow; the long, pale green legs; and the slim and decurved black bill. The species likes to forage in belly-deep water, using its bill to vertically probe the muddy bottom. Stilt Sandpipers are the size and shape of Lesser Yellowlegs but tend to be less hyperactive. Churchill is a well-known breeding ground for the species.

In the gloomy afternoon the tranquility at Goose Creek was disturbed by a dark form passing over the flats—a large raptor on the hunt. It turned, dropped low to the ground, and stooped on a flock of Stilt Sandpipers. The small flock rose into the air and scattered, eluding the big predator—a Peregrine. The majestic falcon landed in a nearby spruce. This individual was in its crisp adult plumage—perhaps the most beautiful of raptors in its adult finery: dorsally blue gray, black hooded with a black "mustache" on each cheek, pearly white throat, and a finely barred breast and belly. Perfection in its avian form. Wouldn't it be remarkable to see this Peregrine hunt down an adult Stilt Sandpiper? That was not to be on this spring evening.

Returning to the Northern Studies Centre as the dinner bell approached, I found the sun was resistant to setting, indicating it was late spring, even though the weather spoke of winter. The birds in the area took due note of the sun and remained active. An adult male

Willow Ptarmigan arrived in the parking lot, posturing and giving his absurd gabbling Donald Duck vocalization. He flew up and posed, midair, and then sailed back to the ground and strutted his stuff and gabbled again. Perhaps most remarkably, he erected his glowing red eyebrow combs, arched above each eye. The sound was weird, but the display was impressive. I didn't see the female that was presumably the unseen target of all this display activity. Then a pair of Pacific Loons raced over the Centre. The loons dropped into an icy pond, and one bird raised its wings to display to the other. It was clearly the mating season in Churchill. The birds were getting a jump on the general thaw of the ponds and the landscape, which would be coming sometime soon. "Spring" is indeed late arriving in a place where the ice buildup on Hudson Bay dominates the local climate.

The next morning broke gloomy with light snow falling. But it was light already at 4:15 a.m. I got out of my narrow bunk bed and gazed out the dorm window at the sub-Arctic scenery: flurries, patches of snow on the open ground, clusters of spruces, and ice-studded ponds in the distance. The spruces, brutalized by the long winter and relatively bare of needles on the windward northern sides of the trees, were doing most of their growing on the lee sides. Redpolls were displaying in the spruces. That encouraged me to head outdoors, into the Centre's "backyard." This is deemed safe for naturalists on the ground because of its proximity to the building's back door.

The spruces hosted not only the perky Redpolls but also singing American Tree Sparrows, White-crowned Sparrows, and the ever-present American Robins; a male robin was giving its familiar and beautiful morning song. This amazing species ranges from my backyard in Maryland (year-round), south to southern Mexico and north to the North Slope of the Canadian Arctic. It is often one of the first birds encountered in the deepest, darkest wilderness of North America. And yet it is also America's backyard bird. Because it is one of North America's most abundant birds, it gets little respect.

Whimbrel *Numenius phaeopus*

Other names: Jack Curlew, Jack, Foolish Curlew, Blue-legs, *Numenius hudsonicus*

The Whimbrel, the prototypical curlew in plumage and behavior, is typically seen in migration stopovers where flocks or small parties forage in a range of coastal or interior habitats.

Appearance: The Whimbrel has a prominent dark eyebrow, crown striping, and a darkish and heavily marked underwing in flight.

Range: The Western Hemisphere populations of this handsome curlew summer in Alaska, the Yukon, the Northwest Territories, and the western shore of Hudson Bay. It winters widely—from the coasts of the southern United States to Tierra del Fuego.

Habitat: North American birds breed on open mossy tundra both in the uplands and wet lowlands. Migrants passing through the Lower 48 frequent sparsely vegetated, higher sections of coastal flats and interior short-grass pastures, as well as a variety of coastal settings, including rice fields, rocky shores, and sandy beaches.

Diet: The Whimbrel's diet is insects, crustaceans, and various coastal berries. It takes the previous season's remnant berries upon first arrival at the breeding grounds, and also takes fresh berries after breeding in late summer. It typically walks and picks prey from the ground, especially favoring fiddler crabs when available.

Behavior: The male carries out a circling vocal display flight over his tundra territory. The species is monogamous. The Whimbrel's song, heard on the breeding grounds, is a musical bubbling series.

Nesting: The nest, situated on a dry rise in lowland or upland tundra, is a depression lined with bits of plant matter. The female lays three or four pale olive eggs spotted with dull brown.

Conservation: The species is uncommon but not threatened. Very large assemblages gather to roost on sandy islands off the coast of the southeastern United States in May. Currently, the North American population is estimated to be 80,000.

Steven Mamet, a soil scientist from the University of Saskatchewan, was my suitemate at the Centre. An expert on the Far North, he was studying the interplay between permafrost and the tree line: the permafrost fosters tundra whereas melting of the permafrost fosters the growth of taiga (boreal forest). One of the advantages of being based at a research center was the access to fellow researchers. They had a lot of knowledge to share. I queried Steve about the details of his research. The future of the tree line and tundra are being impacted by the rapidly warming northern climate—an issue of vital importance to tundra-breeding shorebirds.

On this morning the Goose Creek wetlands hosted 17 Hudsonian Godwits—apparently new birds had arrived overnight from the south. A Wilson's Snipe did a series of winnowing display flights high overhead. Near town three Hudsonian Godwits displayed overhead, swooping about. A Rough-legged Hawk hovered over some tundra. Several seals loafed on ice in the mouth of the Churchill River. Finally, a Silver Fox (actually the "silver" morph of the commonplace Red Fox) showed off his silver-tipped black pelt from an opening in a roadside thicket.

Each morning started with gloom and light snow. On this morning, in the backyard of the Northern Studies Centre, a displaying male Willow Ptarmigan raced all about, landing on the roof and making his silly Disney-character sounds. These ptarmigans were remarkably tame this time of year. A flock of 40 Lapland Longspurs swirled by, behaving like oversized snowflakes. A single Whimbrel passed over the Centre, alerting me to its presence by its high piping notes. This is one of the local breeders in Churchill. The shoreline tundra of western Hudson Bay is favored by the Whimbrel for nesting. This, North America's most widespread curlew, is one of only two of the Magnificent Seven that nest around Churchill—the other being the Hudsonian Godwit. The Whimbrel mainly uses the coastal tundra, whereas the godwit prefers the open boglands that are scattered through the White Spruce woodlands. So, they sort by habitat.

The Whimbrel is Churchill's only curlew, and breeding individuals are shy and skittish, disappearing at the least provocation. During my visit, I saw only Whimbrels in flight, alerted to their presence by their distinctive vocalizations.

Later in the day the sun came out, and at several wet spots I heard calling Boreal Chorus Frogs and Wood Frogs. It is amazing that these amphibians are active and vocal in this harsh season.

The Breeding Season in the Far North

The breeding season for shorebirds in the Far North is brief, normally not much more than six or seven weeks long. Both adults are in a hurry to reproduce and depart southward. The males arrive first, establish territories, and display to the females as they arrive and settle into the habitat. This is followed by nest-building, carried out by both birds (the male selects and fashions the nest cup; the female adds to it). Once the (typically) four eggs are laid, the brooding begins. Godwit eggs take from 22 to 25 days to hatch. The young are up and about the day they hatch, and they can feed themselves within a few days. Depending on the shorebird species, it may be either the male or female that departs the nest site first; for example, Hudsonian Godwit males depart first, whereas Long-billed Dowitcher females depart first. They head to rich staging sites where they will spend weeks feeding and fattening up for the big leap south.

My goal in Churchill was to spend quality time with Hudsonian Godwits on their breeding grounds. This big shorebird nests in the fen (a bog-like wetland fed by mineral-rich waters), south of the Northern Studies Centre. To get to the fen, I needed to negotiate the snowbound Twin Lakes Road. The afternoon sun produced some melting, and a plow had been working on the road. The next morning, Dave Allcorn drove me there, acting as my driver, my gun-toting guide, and my bear-watcher. Dave is a Canadian naturalist based in Churchill and affiliated with the Centre.

I got to spend my last two Churchill mornings in the fen, as Dave was willing to negotiate the snowy road with his nimble Subaru. The fen is tundra-like and boggy surrounded by stands of White Spruce and some Tamarack (American Larch), the latter remaining needle-free at this point in spring. Because it is a fen, it is too wet to be true tundra, but it has many tundra-like qualities: it is open and most of the vegetation is low to the ground—lichens, low recumbent shrubs, some grasses, and some dwarfed trees. It is quite wet, and nearly qualifies as a bog. Thank heavens for my tall rubber Wellington boots—standing water was just about everywhere, and I

An adult male Hudsonian Godwit displays over his breeding habitat.

had to step carefully to avoid deep pools of it. It was cold, foggy, and dreary. And on this morning the shrubby plants were all encrusted with rime ice.

The fen is where godwit expert Nathan Senner started his doctoral research on Hudsonian Godwits as a graduate student at Cornell University. It was a treat to visit "his" Churchill godwits. I would later visit with Senner and his team, in spring 2021, at his postdoctoral godwit research site in Beluga, Alaska, which I recount in Chapter 7.

Within minutes of my arrival at the fen, godwits were making their presence known. Despite the icing and the awful weather, the godwits were gearing up for their breeding season in this prime patch of habitat. (The fen, in fact, figures prominently in the history of godwit field research, for this is where Joseph Hagar did his ground-breaking studies of godwit nesting in the early 1960s, building on the initial Churchill shorebird studies of Joe Jehl and David Parmelee.)

Male godwits were flying over the fen, calling excitedly and swooping like nighthawks. As I saw these birds gracefully zooming about, there was no confusion about their intention—establishing a territory and attracting a mate. When high overhead, the godwits were spectacular. Each was built like a fighter jet with long and narrow pointed wings. The unusual black wing-linings stood out, and the bright white rump patch was distinctive, as was the long narrow bill. Godwits are splendid fliers, and their high-pitched strident cries add to the sexy effect. At one point there were three birds in the sky together. They chased about in the high mist, apparently sorting out the geography of the breeding habitat.

A male descended from the sky and settled atop a White Spruce, his claws grasping the skimpy top branchlets. He was asking to be photographed, and my camera eyepiece filled with the resting male godwit, looking impassively at me from atop the spruce.

Breeding Territory and Establishment

Arctic-breeding shorebirds, godwits included, pair up fresh each spring on the breeding grounds; that is, there are no long-term pair bonds in the migratory shorebirds. Therefore, when an adult male arrives back at his place of birth (or near to it), he must tour the area to find a favorable patch of ground as a nesting territory, and we can assume that there is competitive jostling among the returning males. Success in this competition presumably relates to relative timing of arrival, age of the competing males, physical condition, skill in aerial display abilities, and perhaps even the subsequent intervention by an arriving female, who actively selects one of the competing males or a particular patch of ground that she has found favorable for nesting. That said, it is not uncommon for shorebirds who have bred together to link up again in subsequent years.

A bit later in the fen I saw a male godwit foraging. He was wading in shallow water, probing for invertebrate life in the water, muck, and grassy tussocks with his bill, which has a sensitive tip that can detect prey. The male is a slightly smaller and darker version of the female. The main difference in the breeding season is the color of the breast—very rich red brown with black barring in the male and barred buff and black in the female. This male had been color-banded. It was one of Nathan's birds! The right leg carried an aluminum band and the left a nearly worn-off blue plastic band. It was so cool to see that one of Nathan's birds was still living here. Subsequent emails with Nathan confirmed this was a male that he banded in the spring of 2010. This bird was no less than 10 years old—perhaps considerably older. That is exciting, but not a huge surprise; this big shorebird is expected to have a long life.

The godwits are the prize of the fen, but this is prime nesting habitat for all sorts of waterbirds. Breeding Short-billed Dowitchers,

Stilt Sandpipers, and Least Sandpipers were all on territory. There were also Red-necked Phalaropes, Lesser Yellowlegs, and Wilson's Snipes. Moreover, American Golden-Plovers, Parasitic Jaegers, Sandhill Cranes, Arctic Terns, Northern Shrikes, and Northern Harriers were busy setting up nesting territories in the fen. A morning in the fen was enriched with the sounds of the many bird species that aspire to nest there. It was a veritable symphony of bird voices. Individuals were calling from their terrestrial nest sites, from the shrubbery, from atop the White Spruces, and from high overhead. Hearing all these different voices is one of the joys of being out in this habitat at the beginning of the breeding season.

Vocalizations and Sound Production

Shorebirds produce a wide range of sounds, some on the breeding ground and others during the nonbreeding season, primarily as an alarm or to allow flock mates to communicate with each other. Shorebirds produce both vocalizations and nonvocal sounds. The Long-billed Curlew's song is a long and repeated sweet-trilled whistle. The Upland Sandpiper's flight song is a growling then rising trilled whistle. The Whimbrel's flight song is a series of musical trills or piping notes. Hearing any of these in the late spring is one of the finest surprises for a birder out in the field. Godwits give strident calls. The Willet is perhaps the noisiest of the group, producing incessant vocalizations on territory when it encounters a perceived threat (*pill-will-willit … pill-will-willit … pill-will-willit*).

Similar shorebird species are best identified by their flight calls. The peeps mainly give high-pitched flight calls, key to identification. Our two dowitchers are best identified by their distinctive flight calls.

Up on the Arctic tundra, several of the *Calidris* sandpipers provide a thrilling sound show. The Semipalmated Sandpiper's flight display is a visual and sound treat; its call resembles a quavering electronic tone that carries on throughout the bird's hovering flight. The Red Knot's flight display includes a sad and quavering series of whistles. Wilson's Snipe grows specialized narrow outer tail feathers that can be manipulated during its plunging aerial display flight in spring to produce the ghostly winnowing sounds it is famous for. Finally, the American Woodcock provides a variety of sounds: its wings twitter upon takeoff, a male on the ground in spring display gives a loud *peent* call, and the flight display includes a range of twittering wing sounds with a crescendo when the bird plunges toward the earth.

Spending quality time in prime breeding habitat of the Hudsonian Godwit was a gift. Thank heavens the snowdrifts on the Twin Lakes Road were cleared in time for me to make two early morning visits there. I was starting to get a grip on my target bird's reproductive life. The Churchill sojourn provided a useful lesson about the breeding biology of these birds—mainly the tight seasonal window that the godwits and other shorebirds tuck into in the Far North. A main reason the nesting season of these shorebirds is so brief is that the northern environments are locked in snow and ice well into spring. And on the back end, the birds are in a hurry to get to their staging sites, where they need to invest considerable time to put on the fat and muscle necessary for the long overwater flights southward at the end of summer. Happily, Dave Allcorn never raised his 12-gauge shotgun to ward off a white bear. We stalked the fen in peace.

Midday on June 4 I returned to Winnipeg. After a night in Winnipeg, I started for home. Interstate 29 led me down to Fargo, North Dakota, and then I headed southeast to Saint Paul, Minnesota, then to Milton, Wisconsin, after an exhausting 800-mile drive. My final day was another 800-mile slog to Bethesda, Maryland. I arrived home at nine o'clock, after 15 hours on the road. My trip covered 9,927 miles. ❋

6 High Plains Godwits and Curlews

Our children's children may never see an Upland Plover [now termed Upland Sandpiper] *in the sky or hear its rich notes on the summer air. Its cries are among the most pleasing and remarkable sounds of rural life.*

—Edward Howe Forbush, *Birds of Massachusetts and Other New England States*

THE HUDSONIAN GODWIT, MY MAIN QUARRY, is a sub-Arctic breeder that in spring would lead me into northern Canada and Alaska (see Chapter 7). But its larger cousin, the Marbled Godwit, another member of the Magnificent Seven, is a high plains nester, and it shares breeding habitat with the stunning and elusive Long-billed Curlew, the second largest shorebird on Earth. These two monster shorebirds can be found on the shore in winter, but they retreat to the Great Plains of North America's deep interior to breed. In this vast open country, they are joined by the smaller but equally intriguing Upland Sandpiper (once called the Upland Plover). These are three of my Magnificent Seven. I visited their prime breeding habitat in the springs of 2019 and 2020, and I have combined those two trips in the following narrative.

In late May I drove to Theodore Roosevelt National Park in Medora, North Dakota, crossing the Missouri River along the way. Near Hettinger, I stopped to photograph a Marbled Godwit in a lush, green, grassy wetland. A bit further northwest I stopped again to admire the vista north over the Painted Canyon just east of the national park. The colors of the badlands vary from blue gray to burnt orange. The buttes to the southwest were snowcapped from a late spring storm.

I set up my tent in the Cottonwood Campground in the national park. In the bottom of Cottonwood Creek all the deciduous trees (mainly Eastern Cottonwoods) remained bare of leaves—it still had the look of full winter here in late May. Despite the weather, songbird migrants were passing through. Black-headed Grosbeaks and Chipping Sparrows foraged on the ground beside the park road.

I headed out into the surrounding badlands in search of nesting Long-billed Curlews in the heart of short-grass prairie country. But two hours of back-roads wandering produced no encounters with these stunning birds. Instead, my afternoon's wandering drive through the rolling grasslands passed several prairie dog towns, and I encountered several parties of Pronghorn antelope and a big male Coyote.

A large town of Black-tailed Prairie Dogs sat atop a wind-blown bluff right by the main park road. This was a popular spot for visitors to stop to watch the interactions of these endearing and sociable rodents. I couldn't resist doing this. A few minutes elapsed. Suddenly the attendant prairie dogs cried out in unison as a huge dark shape—an adult Golden Eagle—came up over a rise and cruised low over the colony. The alert rodents dove underground as the hunting eagle passed low to the deck. Yet another failed hunting attempt by a top predator.

The following day I ranged southwestward toward the Montana border in search of curlews on territory. My Xterra bogged down in the gluey red-clay back roads that remained wet from the recent

An adult breeding Marbled Godwit on its wet prairie breeding habitat.

snowstorm. My wheel wells clogged up with this horrible clastic gumbo. There was a lot of nice habitat around but no sign of today's three target prairie species: Sprague's Pipit, Baird's Sparrow, and Long-billed Curlew. These three, plus the Upland Sandpiper, are indicator species of prime prairie habitat. More northbound migrant songbirds arrived in the area, despite the cold and inhospitable weather. Eastern and Western Kingbirds clustered on the verges of the gravel roadbed, apparently hungry and hypothermic. There were no trees for them to perch in.

A third day was devoted to driving the back roads in search of curlews. This was not without its pleasures, because this was picturesque country with grassland hills, badlands, table-topped buttes, and deep gorges. All day long the Western Meadowlark's descending song was the local theme music of the prairie. In the Little Missouri National Grassland south of Golva, North Dakota, the burbling of male Bobolinks announced this species was on territory. But this country appeared shorebird-free. Over the 148 miles I drove, the highlights included visiting scenic Buffalo Gap and Sentinel Butte, and songbirds filled in for the missing shorebirds: Mountain Bluebirds, Baird's Sparrows, Say's Phoebes, and both Black-headed and Rose-breasted Grosbeaks, as well as a couple dozen Pronghorns, Mule Deer, White-tailed Deer, hundreds of Black-tailed Prairie Dogs, and two Coyotes.

At a second prairie dog town in the national park, hundreds of alarm calls signaled an alert that a Coyote was on the prowl. The little rodents do not miss much. Back at the campground, a big old bison bull was hanging around in a patch of cottonwoods near my tent. I photographed him but did not get too close. He was huge and looked ancient. I hoped he would refrain from bedding down next to my tent. These solitary giants move about at will, and there is no telling them where to nap.

Northwest of Theodore Roosevelt National Park lies Glasgow, Montana, and the Bentonite Road, famous to birders in search of short-grass prairie specialists. This is perhaps the finest high plains short-grass prairie country on the continent. The drive from the park to Glasgow was 211 miles and took a bit more than three hours. It was scenic country of dry hills and wide-open cattle-grazing lands. Much of Route 200-S parallels the tracks of a railroad, and there was a 20-mile-long line of boxcars strung out in the thousands on a length of track siding. Boxcars are something of an endangered species on the East Coast. Not so here in the interior.

I camped at the Shady Rest RV Park, which is set on a wooded hill on the verge of Glasgow. My hosts in the campground saw my license plate and told me they were emigrants from College Park, Maryland. The campground was full, but for a fellow Marylander they found me a space to squeeze into. During these COVID days, private RV parks were not filled with vacationers but with down-on-their-luck mobile-home residents, for whom this was the only affordable housing option.

Glasgow itself is a pretty little town of 3,200 set on a railroad line just north of the 134-mile-long Fort Peck Lake, a reservoir of the Missouri River constructed by the Army Corps of Engineers in the 1930s to foster economic development in the Missouri River valley. Glasgow itself is the largest community in eastern Montana. Before European settlers arrived, large nomadic populations of American Bison and Pronghorn in the area supported seasonally mobile Lakota, Dakota, and Nakoda peoples. Through the span of the nineteenth century, the bison herds were brutally reduced to scattered remnants of their former abundance by the advance of the European colonizers.

Here in Glasgow, my field rambles centered on the Bentonite Road, an unpaved and little-traveled tertiary route crossing undeveloped high plains to a bentonite plant. This was literally in the middle of nowhere on lands under the purview of the Bureau of Land Management—the federal agency charged with managing vast expanses of US-government-owned grazing lands through the Interior West. Bentonite is a clay rock produced from volcanic tuffs that has absorbent qualities, making it useful as a drilling mud (in gas and oil extraction), as a component of cat litter, and in a variety of other industrial and pharmaceutical products.

The rough road links Glasgow to the bentonite plant and then continues northwestward as the Skyview Ridge Road, heading toward the tiny rural communities of Hinsdale and Vandalia. The road crosses cattle-grazing country of marginal productivity but is prime high prairie bird habitat. I saw few cattle, but there was abundant fencing, cattle guards on the road, and watering spots. The four nights I spent in Glasgow included daily morning and evening birding rambles out onto the Bentonite Road.

The first couple of miles of the Bentonite Road passed through plowed fields of Glasgow's peripheral farmlands, and these gradually transitioned to sagebrush and prairie. After a few miles more, the countryside opened up with a huge sky and nothing showing but for some distant bluffs and low ridges and the occasional wash. The ground cover was a mix of grasses and low sagebrush, and the colors were muted—buffs, pale gray greens, and various grays. For someone who grew up in the heavily wooded East, this is a different world. Trees were very few and far between—all of them planted as windbreaks by well-meaning ranchers. They survive because they are situated near a water point. This is Big Sky country.

A morning drive started in the chill of predawn—it was May but 37°F. The sky gradually lightened, and the prairie on either side of the road came to life. The air was filled with sound—myriad birdsong from every direction. It was the promise of good things to come, drawing me to the vast open spaces before me.

People think of bird-watching as a visual activity (hence "watching"), but sound is a vital part of the experience (hence the now-preferred term "birding"). My car cast a long shadow across the prairie as I stood beside it, listening to all the birds greeting me. I was not immediately able to recognize many of the songs because the songbirds here were not ones I was familiar with from home. There were Lark Buntings, Chestnut-collared Longspurs, Horned Larks, Vesper Sparrows, and Western Meadowlarks, among others. Horned Larks thronged the roadside; they were extremely common, and their singing produced a continuous loop of sound.

But what of the high plains shorebirds—the grasspipers? Not long after my arrival on the high plains landscape, the swirling musical notes of a Long-billed Curlew were followed by the arrival of a big cinnamon shorebird flying low over the prairie. It put down among some sagebrush and eyed me warily. This was the bird that had eluded me in North Dakota, and here it was, my first shorebird of the day. That long, decurved bill is what sets this bird apart. And the long legs! The vermiculation of the dorsal plumage—black, buff, and brown—helps it to blend into the landscape. This is a bird that has evolved to prosper in these wide-open spaces. The swooping whistles and burbling were electric: here was our largest North American shorebird on its breeding habitat, foraging for insects in the high-elevation short-grass prairie while being serenaded by the vocal songbirds all around.

The big female curlew, with her stupendous bill, stalked the prairie while in the background an Upland Sandpiper—another of the elusive grasspipers—appeared as if by magic. This was the curlew's "little brother." Indeed, the two are relatives. The Uppie's similar plumage patterning helps it hide in these wide-open spaces.

As I slowly drove the dirt track with all the windows down, an insistent and shrill set of repeated notes caught my attention—my car had stirred up a Marbled Godwit, the Hudsonian's big brother. It sailed overhead and landed out in a gravelly wash with very low buff grass and an abundance of seeding dandelion heads, backlit and glowing gray white. Its impressively long bill showed me that it was an adult female. The Marbled Godwit is quite similar in plumage to the Long-billed Curlew but differs in the dark upcurved bill with the pink base. Within a half hour, three of our most desirable grasspipers—curlew, godwit, and Uppie—had made their presence known! That's three of the Magnificent Seven.

A bit further down the road, strange, hooing and popping sounds got my attention. Several bulky birds were moving in the low vegetation on a flat-topped ridge. It was a lek of Sharp-tailed Grouse. Not shorebirds, but grassland fowl doing a remarkable group display. This was the species I had encountered in Glenboro, Manitoba. This lek, on the Bentonite Road, was closer and more active. Several males were challenging one another while the reclusive females scurried around the verges. I could see the males' orange-yellow eyebrow combs, the inflated plum-colored neck pouches, the pointed tails held erect, and the bright white backsides. The Sharp-tailed

Long-billed Curlew ***Numenius americanus***
Other names: Sickle-bill, Old Hen Curlew

The Long-billed Curlew is North America's most magnificent shorebird, the master of the Magnificent Seven. It breeds across the Great Plains and Great Basin, wintering to Central America.

Appearance: The bird has a stunningly long and decurved bill (the female's is the longer). It has cinnamon underwings and a generally rufescent plumage cast, darker on the mantle and palest on the throat and undertail. It is a sister-form to a cluster of three Eurasian curlews.

Range: The Long-billed Curlew summers in sparse numbers across the high plains of the Interior West, from British Columbia to northwesternmost Texas and northeasternmost New Mexico. It winters on the Pacific, Gulf, and Atlantic coasts, the Southwest Borderlands, and in Mexico south to Guatemala.

Habitat: This curlew breeds in dry short-grass and sagebrush prairies of high interior plains, especially near ponds and wetlands. Individuals wintering along the coasts are much easier to locate.

Diet: The Long-billed Curlew's diet is mainly insects; it also takes some crustaceans, small vertebrates, and berries. It mostly harvests these by walking and picking from the ground or surface of the low vegetation. It uses its long bill to nimbly grasp small and elusive invertebrate prey in grasslands and on mudflats.

Behavior: The mongamous male carries out a spectacular vocal display flight over its nesting territory, with a steep ascent followed by a slow vocalizing descent. The voice is reminiscent of the Upland Sandpiper's flight song, with musical and bubbling upward slurs.

Nesting: The nest, set on the ground in open prairie, is a shallow scrape that has a perimeter of plant materials. The female lays three or four clay-colored eggs with dark spotting.

Conservation: The Long-billed Curlew is included on the NABCI Watch List. The species' population is estimated to be 140,000.

Grouse shared this habitat with Greater Sage-Grouse. The larger and darker-colored sage grouse flushed several times by the roadside—females shepherding nestlings through the vegetation. Their lekking season must have already come and gone for the spring.

Killdeers called out from the roadside, but what about the Mountain Plover? The name is absurd, as this species is a bird of plains and salt flats. This was the heart of the northern sector of this species' breeding range. These pale plovers forage on open ground and flats near short-grass expanses. But they are not easy to find. I observed along the road for several days, and the species only revealed itself twice—a threesome in a dry wash and a singleton male lurking at the roadside. They seemed to share this habitat with the considerably more common Killdeer, and it was not clear how the two differed in terms of use of the local habitat. Perhaps the Killdeer is a generalist, whereas the Mountain Plover is a high plains specialist, the former everywhere, and the latter in just a few select areas.

Willets from the western subspecies in a prairie wetland.

Willet ***Tringa semipalmata***

Other names: Humility, Pied-wing Curlew, White-wing, White-winged Curlew, Bill-willie, Pill-willet, *Catoptrophorus semipalmatus*

Appearance: When flushed up from the ground, this large and plain sandpiper transforms itself into a strikingly patterned shorebird because of the black-and-white flashes in the spread-open wings. Birds in the interior west are larger and longer billed.

Range: The Willet summers in two distinct populations—one along the saltmarshes of the Atlantic and Gulf coastlines and the other in the northern Great Plains and Interior West. Both Willet populations come together in southbound migration and on the wintering grounds, and both winter in marshes of the Atlantic, Pacific, and Gulf coasts and southward to coastal South America.

Habitat: The eastern population prefers to summer in coastal saltmarshes. The western birds inhabit interior wetlands.

Diet: The species' foraging style is diverse: pecking, probing, and chasing active prey in water. Its diet is insects, crustaceans, other small invertebrates, and fish fry. The interior birds tend to consume more arthropods and the eastern coastal birds more crabs.

Behavior: The Willet is noisy on the breeding grounds, where it tends to nest in colonies. When disturbed by an approaching birder, it will fly about giving its *pill-will-willit* call and will bob its body in the manner of a yellowlegs. The species is monogamous.

Nesting: The nest, set on the ground in thick marsh grass, is a shallow depression lined with living and dead grass. The female lays three or four clay-colored eggs thickly spotted with chocolate brown, mainly at the larger end. Both adults incubate the eggs. Parents lead the young to a productive waterside foraging site. The female departs first, leaving parental duties to the male.

Conservation: The Willet is categorized on the IUCN Red List as Least Concern, but it is included on the NABCI Watch List. The species was gravely reduced by market hunting in the late 1800s and early 1900s, but today's population is 250,000.

The most desired shorebirds on the Bentonite Road were members of the Magnificent Seven, but, in fact, the most common shorebirds on the Bentonite Road were Wilson's Snipe, Wilson's Phalarope, and Willet. The western (interior) population of the Willet tends to hang out near water features, especially prairie ponds—particularly those with a fringe of marsh grass. In this prairie country, Willets are the most vocal of the resident shorebirds—just like they are back in the saltmarshes of the East. The species is slightly smaller than the Hudsonian Godwit but the two look similar when seen from a distance. Willet hatchlings wander from the nest within a day of breaking the egg. They can feed themselves at this point. On the Bentonite Road, singletons of the species nervously worked the edges of the scattered roadside ponds.

But what about the songbirds and their all-encompassing symphony of morning song? Yes, this was great shorebird country, but the "dickey birds" dominated. Because there were no trees about, songbirds of the high plains announced their presence with song flights. My first encounter was with a male Lark Bunting, rising into the sky, belting out his chattery musical song, then parachuting to the ground with wings outstretched to show off the white wing slash contrasting with the all-black body. The nearby female, the focus of the male's activity, perched on a low sagebrush. She was drab, streaked, and sparrow-like, revealing the species' genealogical roots.

The Horned Lark (unrelated to the Lark Bunting despite the shared name) did an even more dramatic display flight, ascending high overhead and then circling about and singing its high tinkling song. Its black "horns," black tear drops, and black breastband set against the yellow throat made this lark perhaps the prettiest songbird of the high plains. Moreover, this bird was no doubt the most abundant bird species of the Bentonite Road: as the car moved slowly down the road, lark after lark flushed up from the roadbed, so many that it was impossible to keep count. It was gratifying to see a beautiful bird in such great abundance.

Longspurs breed there too—both Chestnut-collared and Thick-billed (formerly McCown's). The adult males of both are handsome, but in different ways. The first features an all-black breast and chestnut nape; the second features a black breast bar and a chestnut wing bar. Both are vocal in spring and primed for song flights. At this time of the year, males were chasing females, and males were chasing males, and twos and threes perched in the middle of the dirt road, sorting things out, mano a mano.

The best songbird surprise was Sprague's Pipit performing its song flight. This plain little songbird is one of the most elusive birds of the short-grass prairie. I have worked many a prairie for this pipit without success. On this day one launched out of the grass up into the sky, flaunting its white tail feathers. Like the Horned Lark, the pipit gave its musical and descending series of notes from on high; the voice was more prominent than the bird itself—a descending series of musical tinklings unlike any other prairie songster. A displaying Sprague's Pipit was a sure sign this was some of the best high plains habitat to be found in Montana.

A remarkable thing about this wide-open and treeless habitat was that it could hide numbers of mammals, even large ones. Individual male Pronghorns, as well as parties of females and young, appeared and disappeared in the subtle contours of the landscape. Mule Deer parties did the same. One minute nothing, and the next minute—there they were! And raptors too. Overhead a handsome juvenile Golden Eagle soared by, flashing its white tail patch, and a Northern Harrier quartered low over the ground in search of some unwary small mammal. A Swainson's Hawk passed by. And lots of ducks loafed in pairs on the ponds and wetlands that were scattered here and there. Unexpectedly, in a line of planted cottonwoods, a pair of Red-headed Woodpeckers was nesting in a dead stub, making it evident they only required a handful of trees to establish a breeding territory. With trees scarce, these woodpeckers must rely on insects captured on the wing to survive and to feed nestlings.

Upland Sandpiper ***Bartramia longicauda***
Other names: Uppie, Upland Plover, Highland Plover, Land Plover, Grass Plover, Hill-bird, Pasture Plover, Prairie Snipe, Quailly

Looking like a small, short-billed curlew, this elusive denizen of interior prairie habitats is typically seen as a wary singleton or in pairs. The "Uppie" is indeed a close relative of the curlews, and its bubbling flight call is quite curlew-like. When it lands upon a fencepost in a rural field, it briefly poses with its wings held up.

Appearance: The bird's mantle is heavily vermiculated with browns and black, its throat is finely dark streaked, and its belly is white. Its large, dark eye is distinctive, and it has a longish heavily marked tail.

Range: The Uppie is an uncommon breeder in prairie and pasture from Alaska and the Northwest Territories southeast to Kansas and Maine. It is patchy and thinly distributed in the East. The species winters in the South American Pampas of Argentina and Uruguay.

Habitat: The Upland Sandpiper inhabits prairies or prairie-like habitats, where grass does not exceed eight inches and where there are patches of open ground. In migration, it can be found at sod farms, bare fields, and expansive mowed grassy areas of airports.

Diet: The Uppie hunts for terrestrial invertebrates, mainly arthropods, which it captures using a run-and-grab gait. It will take waste grain on occasion. It is usually seen foraging singly or in pairs.

Behavior: The species is monogamous. The male carries out a fluttering and vocal display flight over his territory. The bird walks steadily with its head moving back and forth. The Upland's song is a rising musical trill, given by the male in flight.

Nesting: The nest is a shallow scrape on ground, lined with dry grass. The nest is usually hidden by a covering canopy of grass. The female lays four creamy-buff eggs spotted with brown.

Conservation: Market and sport hunting in the late 1800s led to the decimation of this species in the Great Plains. Today the species population is estimated to be 750,000 and stable.

A pair of Upland Sandpipers was apparently nesting in the grassland near the woodpecker nest tree. The two Uppies moved through the low grass and made complaining scolds at my presence. Their small heads poked up from the safety of the grass. As they moved, their heads drew back and forth in synchrony with their syncopated gait. The paired adults kept together as they monitored my movements. Then one individual flew up onto a fencepost and raised its wings momentarily before slowly closing them over its back.

From Glasgow, I headed southeast to Glendive, Montana, for a high plains adventure of another sort. I set up my tent and tarp in Green Valley Campground, with plans to visit Makoshika State Park and to spend a morning with Shana Baisch on her cattle ranch. The state park features some spectacular badlands and good birding habitat. The Baisch ranch features excellent exposures of the Late Cretaceous Hell Creek Formation in a set of badlands, with an abundance of dinosaur bones weathering out of those sediments. Shana hosts fossil hunts in summer—"Baisch's Dinosaur Digs."

Makoshika State Park's 11,000 acres include stunning badlands scenery and camping and hiking. In the early morning, I birded the lower section of the park and enjoyed close encounters with a singing male Lazuli Bunting (cerulean, white, and rufous), a female Bullock's Oriole (mainly orange), and a Spotted Towhee (black, white, and red brown). A pair of Great Blue Herons played about high overhead, like two slow-moving aircraft trying to wax each other's tail. I couldn't tell whether these were two males in a territorial battle or a male and female courting.

At the Baisch ranch, Shana greeted me and then led me by SUV out into her badlands, where it was good to have a four-wheel-drive vehicle with high clearance—the "road" was only a faint trace across the rough-hewn Montana landscape. Welcome to Hell Creek!

The first stop was marked by a huge blue tarpaulin. Pulling back the covering, Shana showed me her current excavation—the skeleton of a *Triceratops* that Shana had identified as the species

T. horridus. It had been slowly weathering out of a flat hilltop. Various major bones were poking up out of the gray-brown earth. Excavating a skeleton the size of a *Triceratops* is a substantial enterprise, and this one would require months to complete.

We spent a couple of hours clambering up and down the bare-earth badlands. Every few minutes, Shana called me over to point out a fossil bone weathering out of the sediment—some larger, some smaller. There was a rib bone (possibly of a *Triceratops*). There was the tip of a small bone protruding from the sandy hardpan. She handed me small bits of dinosaur bone, explaining how she could identify it as bone. She handed me a tiny item with a strange shape and blade-like edge to one side—the tooth of a *Triceratops*.

Together we came across more than 50 bits of what were once living dinosaurs. This was an immensely productive site—I guess qualifying as a "bone bed." One of the main takeaways was that the moment a dinosaur bone makes it to the surface, its inevitable destruction begins. It is thus a race against time to harvest these fossil treasures before sun, rain, and erosion do their dirty work.

A lot has been written about the conflicts between private commercial collectors and academic researchers and the future of dinosaur fossils. This conflict is exemplified by the drawn-out legal battle over the proper ownership of the *Tyrannosaurus rex* skeleton named Sue, collected from a site in Faith, South Dakota, in August 1990. Private collectors have been described as enemies of paleontology. In reality, there are not nearly enough people out there collecting and preserving the fossil skeletal material that is being laid bare by erosion. And, of course, properly excavating a dino skeleton is a huge amount of work, as Shana was learning firsthand. State and federal governments need to craft laws and regulations that foster the recovery of these ancient relics. Having both private for-profit as well as public nonprofit organizations doing this will result in more of these fossils being saved. The laws need to be more friendly to both the private collectors and the professional paleontologists.

Regulations should promote the conservation of the fossils through the engagement of as many players as possible, all fairly regulated.

Aside from *Triceratops*, Shana's property yields good numbers of bones of *Edmontosaurus*, one of the duck-billed dinosaurs (a hadrosaur). And this is a Late Cretaceous paleoenvironment that also harbored the great *Tyrannosaurus rex*. Shana found a giant incisor of this species on a hillside the previous summer—one of the more exciting pickups of her career. She pointed to where the gorgeous *T. rex* tooth had been lying right on the surface, awaiting discovery. She had marked the site with a small piece of tinfoil. A few feet above this site was a thin black band of sediments that Shana thinks marks the end-Cretaceous extinction event that heralded the demise of the dinosaurs some 66 million years ago, when an asteroid struck the Yucatán Peninsula, creating a global cataclysm. It was entrancing to dig out some of that black sediment from the cliff on Shana's ranch—a product of one of the major turning points in Earth history. After we returned to her house, Shana brought out a small wooden box and opened it to show me that glossy brown six-inch *T. rex* tooth, with its serrations still sharp. What an amazing artifact to hold!

What does any of this have to do with a hunt for living shorebirds? Well, Shana showed me fossils of an ancient lineage that gave rise to the class Aves (the birds). Dinosaurs and birds? Just look at the footprint of a cassowary from Australia and see how it resembles the track of a small dinosaur. In Glendive I was paying homage to the godwits' ancestral roots. A decent way to spend a spring morning.

While I was at the Baisch ranch hunting fossils, a strong prairie wind trashed my camp. It blew plates and dishes off the picnic table, knocked over my bicycle, and ripped my big tarp out of the ground, sending it sprawling across the campsite. It reminded me that the high plains are a land riven by winds. The tarp acted just like a sail. I had not tied it down well enough to bear the force of the gusts of wind. The next day, when I stopped in Richardson, North Dakota, on my way east, another strong gust nearly blew off my car door.

Recent eBird reports of godwits drew me eastward through North Dakota. South of Bismarck, I set camp at Fort Abraham Lincoln State Park, in the town of Mandan, right on a bend of the Missouri River. The fort had been commanded by George Custer before his resounding defeat at the Little Bighorn. An eBird report led me to McKenzie Slough, which was birdy but free of any godwits. Instead, the site offered up stilts, avocets, and Stilt Sandpipers.

A long drive eastward got me to Fargo, North Dakota, where it was sunny and 80°F. I camped in Lindenwood Park, in downtown Fargo, my tent situated a mere 75 yards from Interstate 94. This was true urban camping! I somehow managed to sleep, mainly because the night's interstate traffic was light. Early the next morning I was off to the "Shrike Unit" of Felton Prairie State Natural Area, in northwestern Minnesota, hunting for Greater Prairie Chickens and Marbled Godwits. No luck with the prairie chickens (perhaps their display season was over), but a mated pair of Marbled Godwits on territory entertained me with display flights and calling bouts from this nice remnant of tallgrass prairie. Dozens of Sedge Wrens, Bobolinks, and Western Meadowlarks serenaded me, and a Wilson's Snipe winnowed high overhead.

As Memorial Day approached, I camped at Greenwood Lake, north of Duluth, on Route 2 in Minnesota, as I made my way circuitously toward home. This was a favorite boreal forest hotspot. I was taking a short break from shorebirds to check out some boreal songbirds, because we were approaching the arrival peak of the migrating songbirds in the north country (apparently aligned with the emergence of the biting insects). Camped in an old logging clearing, I spent a lot of time biking on Route 2, flat as a pancake and traffic-free. This is one of the finest spots for dawn bicycle birding in North America. In 24 hours I logged 16 species of wood warblers plus a bunch of boreal specialities: Long-eared Owl, Black-backed Woodpecker, Canada Jay, Merlin, Yellow-bellied Flycatcher, and Pine Siskin. The needle buds of the Tamarack trees were just starting to "break." It was still wintry looking up there. As it warmed up, the first of the black flies started to home in on my camp. Northern Cricket Frogs were beginning to vocalize. My only shorebirds were displaying Wilson's Snipes at dusk and Spotted Sandpipers working the lakeside gravel in the chill of the sun-blessed early morning.

From Greenwood Lake I move south to Sax-Zim Bog for more boreal magic. I set camp on the banks of the Little Whiteface River in a wooded clearing. I was camping at the dead end of a dirt road because I could not find an open campsite (COVID was still raging, and campsites were shut tight). I slept inside my car, drifting off to sleep listening to the soft rumble of distant freight trains.

The following day I surveyed Sax-Zim Bog, the most famous birding locale in Minnesota. Not a lot of other birders were out and about (due to COVID), but I bumped into Mike Hurben and Claire Strohmeyer, who were there to hunt for LeConte's Sparrow. We tramped out into the wet meadow south of Stone Lake Road and chased after this elusive little sparrow in its well-known breeding spot. Forty minutes later we had our cameras focused on the bird. The Hurbens were happy, as was I, with excellent images of this cunning little sparrow, but we could not celebrate with high fives because of the strictures of COVID social distancing.

I could hear the whisper of the sparrow's minimalist vocalization that morning because I was now wearing hearing aids, which allow me to detect high-pitched songs. When I was last in northern Minnesota with my wife, Carol, she was continually pointing out a sound and asking what bird it was. I could only say, "I don't know, darling—I can't hear it!" Hearing aids are a miracle for aging birders.

I had one more morning at the boglands, having spent the night again sleeping in my car, this time in the parking lot of the Sax-Zim Welcome Center on Owl Avenue. While loading up the car at dawn, I heard a pack of wolves howling nearby—that's a Minnesota highlight! Another bonus was a singing male Connecticut Warbler. According to eBird, this was the first Connecticut of the season here in Sax-Zim. I burrowed into a wet thicket to photograph the singing bird, but the bird had other plans. No photograph.

Driving northeast onto the Upper Peninsula of Michigan, I was led by two eBird reports of Hudsonian Godwits. Both proved to be time-consuming dry holes. But I did get a firsthand experience with the effect of Lake Superior on local weather. The interior of the Upper Peninsula was 81°F and toasty, but at the shore of the lake in Baraga Township, it dropped to 57°F. That's some seriously chilly water!

Late in the afternoon, with no campground in sight, I put up in a seedy motel in Munising, Michigan. Because of the COVID emergency I was required to complete a set of declarations stating that I was on official business travel to gain access to the motel room, which, let me tell you, was nothing special.

Traveling south across the Mackinac Bridge and down the Lower Peninsula of Michigan to Grayling, I queried eBird to find a nice youthful plantation of Jack Pines where I could photograph Kirtland's Warbler. Once I was in the monoculture plantation of low pines, it was simply a matter of listening for the loud and chattering song of the males (*chip-chip-weeda-weeda-weedit*) on territory. The warbler is famously tame and unwary. Several times I found myself backing up to get enough distance so my lens could focus down on the little bird. This stand of pines had half a dozen Kirtland's Warblers nesting in it; it was like shooting fish in a barrel.

Next stop: the famous spring migrant songbird hotspot Magee Marsh, on the southern shore of Lake Erie near Port Clinton, Ohio. My first stop was the Metzger Marsh shore woodlot. Highlights included a Philadelphia Vireo with its yellow-washed flanks and another new bird for my trip list—a male Black-throated Blue Warbler. Next, I drove to Magee Marsh. The road was blocked—no entrance due to COVID! I retired to the Riverview Campground for one last night on the road. A peaceful sleep readied me for the eight-hour drive home. During the journey home, I added the Indigo Bunting to bring my trip list to 195 species. ❋

7 Beluga Godwits

The wild inspiring sky song of the dowitcher, as it cruises aloft on swift saber pinions, [is] *one of the most unforgettable of Arctic voices. All this, too, in a land where the very air is vibrant with bird cries and melodies of various plumed minstrels.*

—Herbert Brandt, quoted in Henry Marion Hall, *A Gathering of Shorebirds*

IN MID-MAY 2021 I FLEW TO SOUTHWESTERN ALASKA to spend time with Hudsonian Godwits nesting on their bogland (muskeg) breeding habitat. This field trip had been arranged with shorebird expert Nathan Senner, then a professor at the University of South Carolina who was doing long-term field studies on Hudsonian Godwits in Beluga, just west of Anchorage. I shadowed Nathan's team to take in the details of his field work and learn more of the godwit's nesting behavior. Recall that back in Churchill, Manitoba, I had encountered a godwit that Nathan had color-banded in 2010. So now I got to spend quality time with the world authority on the Hudsonian Godwit at his new study site.

Arriving in Alaska ahead of the Senner team, I spent a couple of days in Anchorage to settle into the new environment. Local birding expert David Sonneborn took me out birding a couple of times, and we managed to see Hudsonian Godwits in the mudflats of Cook Inlet in Anchorage harbor. The godwits had already arrived for the spring. Anchorage is a stunning venue for those who love the out-of-doors—fjord-like coastal waters, snowcapped mountains in every direction, and ready access to birds that are exotic to those of us who live in the Lower 48. The Arctic Tern, for instance, is a commonplace and confiding breeding bird on ponds and wetlands right in the city proper, as is the Red-necked Grebe. Hudsonian Godwits and Whimbrels breed in habitats not far from Anchorage.

One afternoon I visited Arctic Valley Ski Area in the Chugach Mountains, just a few miles, as the raven flies, east of downtown. The road to Arctic Valley gave ready access to the mountainous uplands and rocky tundra. I hiked up to the top of a mountain ridge next to a former Nike-Hercules missile site—ringed by forbidding "Stay Out!" signs—from the early 1960s. I walked the upland tundra in search of Rock Ptarmigan. Recall that I had bumped into the more commonplace Willow Ptarmigan several times on my visit to Churchill. The Rock Ptarmigan was a more difficult bird to track down. I hiked the ridgetop tundra for several hours without seeing my bird. Then I did a sound playback of the bird using my iPhone, and suddenly three all-white adult Rock Ptarmigans appeared out of nowhere, flying and vocalizing.

One individual perched on a rock pinnacle overlooking the waters of Turnagain Arm. I cautiously approached this grouse with its dark red eyebrow comb and all-white winter plumage until I was too close for my zoom lens. I then shifted to my iPhone and got full-frame images that I could email to family and friends. This confiding bird never flushed, no matter how close I approached. I left the bird where it was perched and headed back down to the car. On my hike down, I found Arctic Ground Squirrels and Golden-crowned Sparrows. These uplands offered stunning vistas of the Cook Inlet and the snowcapped Chugach Range to the east and the snowy

Alaska Range to the west. For a first-time visiting naturalist, western Alaska is over-the-top amazing.

To get to Beluga I chartered one of Spernak Airways' single-engine Cessnas out of Merrill Field, the downtown airport where the bush charters take off to fly wide-eyed visitors to wilderness spots in the interior. Because roads are few, bush planes are still an important means of transportation to the backcountry for fishing, hunting, and getting to isolated Indigenous communities. My target airstrip, Beluga, services an Indigenous community named Tyonek as well as a remote power plant that generates electricity for Wasilla and other northern suburbs of Anchorage (and, no, you cannot see Russia from any of these places). Tyonek is home to an Athabascan community speaking the Dena'ina dialect.

As this was my first time in Alaska, I really had no idea what to expect in terms of lodging, food, transportation, and access to good godwit habitat. As it so happened, things played out as smoothly as my 20-minute charter flight from convenient Merrill Field. Weather was good for the flight, and from aloft, vistas to the northeast featured the snowy Alaska Range as well as the great snow dome of Denali—at 20,310 feet, the highest summit in North America. There was lots of water below—the Cook Inlet mainly, but also the Knik Arm reaching northeastward and the Turnagain Arm stretching southeastward. The mighty Cook Inlet stretched southwestward to join the Gulf of Alaska. We flew mainly over undeveloped land that supports a mix of bog/muskeg, spruce monoculture, and, on higher and drier places, stands of aspen, whose leaves were just beginning to break from the bud. Below were flocks of Snow Geese, their plumage white against the dull colors of the wetland. The waterlogged boglands were dun colored and broken up by myriad ponds and marshlands. This was Alaska as imagined—undeveloped and waterlogged land stretching to the horizon.

After the pilot dropped his plane down onto the long gravel runway, Max Vigeant, one of my hosts, greeted me and drove me to the lodge at Beluga Fish Camp, getting me checked in to a rustic but capacious bedroom suite plus arranging a car for driving the local gravel roads. Max and his colleague, William Fredette, run the Fish Camp, which is set in a clearing on a high bluff overlooking Cook Inlet. The lodge mainly caters to guests from the Lower 48 who visit for the seasonal runs of Coho, Sockeye, Chinook, and Pink Salmon, which ascend the nearby rivers from mid-May to late September. Off season, most of the Fish Camp's business comes from local clientele who favor the Camp's well-stocked bar. As their sole overnight guest, I got great service—three nice hot meals a day, prepared just for me. The boglands that support nesting Hudsonian Godwits were just a few miles from the lodge, and the road system gave access to a whole array of upland and adjacent coastal habitats—ideal for birding and enjoying nature.

I spent a week based at the Fish Camp, focusing on the Nathan Senner team and the nesting godwits in the north and south bogs. In my spare moments, I wandered all about in search of birds and other wildlife. Much of the landscape is truly wild, but the Beluga area has been developed because of a large gas-fired power plant just north of the airstrip. The plant is situated there because there are producing gas wells right there, supplying all the needed energy to make the generators run. This facility explains why there are roads, power lines, and settlements other than Tyonek village. There's even a little Beluga General Store, crammed with all manner of supplies, snacks, and drinks. I was not exactly roughing it in rural Alaska! There was even good internet and phone access—a nice mix of civilization's amenities with ready access to wild lands filled with birds and mammals.

The boglands were indeed boggy—they were very wet. Open water mixed with inundated boggy grasslands. This was not tundra but rather a true bog. Spruces managed to take hold on any ground that rose above the standing water. These higher areas also attracted a range of woody shrubs. Thus there was a nice mix of vegetation that offered a range of nesting habitat for birds. My first morning was spent in the north bog. Nathan and his team had not yet arrived in Beluga, so I was on my own. This gave me a chance to get a sense of the habitat where they would be working. Donning my tall rubber boots, I waded out to one of the small spruce-clad hillocks in the bog and sat down to watch and listen. The air was still, and frost lingered on any vegetation that remained shaded from the rising sun.

The bog was, more than anything else in spring, a gull breeding colony, with many Short-billed Gulls, quite a few Bonaparte's Gulls, and some Glaucous-winged Gulls in attendance. The gulls milled around over the bog and made a lot of noise. The very aggressive Bonaparte's Gulls dive-bombed me regularly. Pairs of Arctic Terns were setting up nesting territories and building nests. There were nesting shorebirds too—Least Sandpipers, Wilson's Snipes, Short-billed Dowitchers, Pectoral Sandpipers, and Hudsonian Godwits.

It did not take me long to hear the godwits' high-pitched complaints. Males circled overhead. When taking a break from their overflights, they perched precariously atop the White Spruces (just as they did in Churchill). I saw a female godwit foraging at the edge of a pool, and after a while it became evident that there were two male-female pairs moving about in my area. Hidden by the spruces, I happily monitored these godwits, who were foraging intently out on their nesting territories in this Alaskan wetland.

Diet on the Nesting Grounds

Arctic and sub-Arctic breeding shorebirds shift diets when they move from their nonbreeding habitats to their nesting environments. For most of the year the birds feed on a broad array of invertebrates, including worms, mollusks, crustaceans, and other tiny prey found in marine and estuarine wetland habitats. By contrast, on the tundra or fen or bog, the birds encounter a much more restricted menu—mainly terrestrial arthropods found in the vegetation or aquatic invertebrates. In some cases this is balanced by marine invertebrates

The Dowitchers
1. Short-billed Dowitcher *Limnodromus griseus*
2. Long-billed Dowitcher *Limnodromus scolopaceus*
Other names: Red-breasted Snipe, Robin Snipe, Kelp Plover, Deutcher, Fool Plover, German Snipe, White-tailed Dowitcher

Appearance: These two dumpy, long-billed sandpipers look like a cross between a snipe and a godwit. The two dowitcher species are essentially impossible to distinguish by plumage or bill length—despite their names. The two were recognized as distinct by Rowan in 1932 and Pitelka in 1950.

Range: The Long-billed breeds in northeastern Siberia, northern and northwestern Alaska, and northwesternmost Canada. The Short-billed breeds from southwestern Alaska eastward to Manitoba and Labrador. Both species winter to both coasts. The Long-billed winters south to Mexico and the Caribbean, and the Short-billed winters as far as South America.

Habitat: When nesting, the Long-billed uses tundra and the Short-billed uses open bogs in spruce forests. In migration, the Short-billed prefers coastal saltwater habitats, whereas the Long-billed prefers interior freshwater habitats.

Diet: The two species feed upon arthropods on the breeding grounds and benthic invertebrates (which inhabit the sea or bay bottom) on the wintering grounds. Both will, at times, take plant matter of various kinds.

Behavior: The two species can be distinguished by their flight calls. The Long-billed's is a *keek* and the Short-billed's is a *tuu-tuu-tuu*.

Nesting: The nest is a depression in low open vegetation. The female lays four olive or buff eggs marked with dark brown. Both adults incubate the eggs, but the male mainly cares for the nestlings.

Conservation: Both populations have declined, likely for a variety of reasons, including habitat loss and climate change. The Short-billed is Watch Listed (population 150,000). The healthier Long-billed's population is more than 500,000.

if the nest site is within a short flight of a bay or estuary where the adults can feast on their standard nonbreeding fare (as was the case in Beluga because of the proximity of the Cook Inlet mudflats). That said, the young of the year must subsist on arthropods they locate on their own in the low tundra or bog duff around their nest.

The nesting Short-billed Dowitchers forage in the wettest sectors of the boglands along with the godwits. These could be likened to a "poor man's godwit"—not as rare, not as tall, and not quite as handsome as the adult male Hudsonian Godwit in its spring breeding plumage. Spring plumage of the dowitcher is godwit-like, with the rich buffs and rusty browns and the dark flecking on the back and flanks. The bill is long and impressive but not recurved. The dowitchers feed in deep water but are not quite as long legged as the godwit.

Just to the north of the bogs rose the snow-covered Alaska Range. Sleeping Lady, the snow-bedecked mountain with the alluring curves, was prominent in the near distance. In the boglands, spring was just starting to announce itself through the clamoring birdlife. White-crowned and Lincoln's Sparrows sang from the thickets. The gulls continued to make a racket. The deciduous vegetation lagged behind, as always, slow to bud out. A pair of Red-throated Loons passed low overhead. Tule White-fronted Geese flew over in pairs, as did Sandhill Cranes.

This was the 12th year Senner had been working on godwits. Because of the COVID-19 pandemic, this spring Nathan had a skeleton crew working in Beluga—Nathan, research biologist Maria Stager (Nathan's wife), and research assistant Lauren Puleo. Normally, Nathan would have a larger team of students covering a broader palette of field activities.

Nathan generously allowed me to shadow him and his team for several days. This was about as good as it gets when it comes to establishing an up-close-and-personal engagement with my focal species. Nathan's operation that year was straightforward: (1) locate adult godwits on territory, (2) find active nests, (3) using a mist net, safely capture males and females while they are on the nest, and (4) gather biometric data on each bird and attach color tags and geolocators as needed. The team had a goal of finding 10 nests in the south bog and 20 nests in the north bog. This would be seriously challenging work, as godwits hide their nests fiendishly well, and they do not flush readily from the nest.

By watching the activities of each pair of godwits, the team gained a sense of where a nest might be generally situated. Then it was a matter of marching, side by side and back and forth, across the bog until a sitting bird flushed. The moment the bird flushed, there was a careful rush to the point where the bird emerged from the grass. If things went well, the nest, with four dark-blotched greenish-olive eggs, was found. A marker stick was placed near the nest site—always the same cardinal direction from the nest and far enough from the nest to avoid attracting the attention of nest predators (the Common Raven, Short-billed Gull, Sandhill Crane, and Northern Harrier), which learn to cue on such things. Senner told me it requires 24 person-hours of effort to find each nest.

Each nest was hidden in low bog vegetation—a mix of grass and recumbent shrubs—on a small rise that put the sitting adult and the eggs above the waterline. The rise was barely perceptible, but it was crucial, as eggs in a flooded nest do not survive. The nest itself was a scanty depression below the vegetation line. When the adult hunkered down on the eggs, her back and head lay well below the height of the surrounding low scrub. Thanks to the adult's vermiculated dorsal plumage, the sitting bird was invisible among the dully colored bog vegetation.

One day we quietly approached a nest located the preceding afternoon, marked by its telltale stick. Maria pointed at the nest, just a few feet from us. Staring hard at the spot she was pointing to, I saw no sign of adult or nest. As I crept closer and closer, the head and bill of the adult, motionless over the eggs, became visible, then I saw the patterned back. This is camoflage to the max.

More on Nesting

After pairing, northern shorebirds construct their nest by creating a shallow depression on open ground or low tundra or bog vegetation. The nest cup is lightly lined with soft vegetation and the depression is just big enough to hold the expected four eggs. Shorebird eggs are pear-shaped, with distinctly tapered smaller ends. This allows the eggs to be neatly situated in the nest with the narrow ends toward the center, presumably making brooding more efficient, allowing all the eggs to get a similar amount of heating from the sitting adult. Most eggs have a pastel background color of olive, buff, reddish, or purplish. The broader end of the egg is blotched with marks of dark brown, and the narrower end is scrawled with dark squiggly pen lines. This renders the eggs cryptic and difficult to be seen by a predator such as a raven or a jaeger if the adult is off the nest.

Tundra-nesting shorebirds can be remarkably tolerant of human approach when nesting. When closely approached by an observer, the adult Hudsonian Godwit on the nest will simply hunker down flat and remain motionless. The dorsal plumage of many species renders the birds and their nest virtually invisible to those passing by. Some species, such as the Long-billed Dowitcher, remain on the nest until the last moment, and then the sitting adult will noisily explode off the nest to surprise and confuse the intruder. A tundra-breeding Ruddy Turnstone, on the other hand, will leave the nest and then produce alarm calls to draw the intruder away. The Purple Sandpiper is masterful at a similar strategy that has been named the "rodent run"; when the potential predator approaches to within a yard or two, the sitting adult will jump up and race away in a zigzag fashion, squeaking like a mammal and erecting its dorsal feathers to look like fluffed-up fur. This presumably works because lemmings and other small rodents are favored prey of northern predators. Some nesting parent shorebirds, including Red Knots, curlews, godwits, and shanks, launch out to attack overflying aerial predators such as jaegers, raptors, and ravens.

Once the team located and marked the godwit nest, its next objective was to capture both the female and the male when they were sitting on the eggs. Each parent broods the eggs for long bouts; the female usually sits during the day, and the male sits during the night. I assisted the team with the location and capture of a female on the nest. This was a rather simple process, relying on the parent's tendency to "sit tight" and not flush until an intruder gets very close. The team strung up a mist net between two poles, and two people held it taut and upright. These two then slowly walked up to the nest and ever-so-slowly lay the net down flat over the nest. At that point Nathan approached the nest and gently grabbed the bird, which had flushed up into the now-loose netting. Nathan gingerly extracted the bird from the netting and placed it in a cloth holding bag.

If the bird was unmarked, it received a US Fish & Wildlife Service metal band as well as a color-coded set of plastic bands on the lower legs, allowing the bird to be identified from a distance. Then the bird was weighed, and several body and bill measurements were taken. A small sample of blood was collected to provide data on hormone levels as well as DNA for paternity studies. Finally, the team affixed a geolocator (a data-logger) to the bird's upper leg, above the leg joint. By keeping track of sunrise and sunset, this tiny device records the bird's travels south for the winter and back north the next spring. Eventually, Nathan would have to recapture the bird to remove and download the contents of the geolocator to find out just where the bird had wintered and by what route it left and returned to Beluga. Geolocators are simple and inexpensive and provide the general details on the large-scale annual travels of the migrating birds.

While we were capturing the sitting female, its mate was likely out on the flats of Cook Inlet, foraging for marine worms, tiny bivalves, and other mud-living saltwater invertebrates. Cook Inlet's tide is a monster—26 vertical feet, twice a day. The dropping tide unveils a mile-wide set of flats rich with food for godwits and other shorebirds. The flats are visible from the tall bluff at the edge of the big lawn at the Fish Camp. A scattering of shorebirds, mainly Hudsonian Godwits and Whimbrels, often move across the glistening mud. They work slowly and carefully, poking and probing and swallowing. It is this expansive and rich feeding site that makes the north and south bogs so attractive to nesting godwits. Their flight from bog to feeding flats is no more than a mile. Each member of the pair can spend hours out at the flats each day, thanks to the super-long days in Alaska in late spring and early summer. Long sunlit hours mean lots of feeding time and plenty of nutrition to keep the godwit adults going, and then to initiate the "fattening up" that they will continue in a nearby staging area prior to their big flight south.

Precociality and Parental Care

Shorebird clutches are brooded for 19–31 days, and then the clutch mates pip their eggs synchronously. The unhatched young have an egg tooth sitting atop the upper mandible that they use to break the egg open. This falls off shortly after hatching. The sitting parent immediately carries away the eggshell remnants, which might otherwise reveal the nest location to predators. Once the eggs hatch, the chicks are quickly out and about, in some cases foraging for invertebrates in the bog vegetation the very day they hatch. These chicks are precocial, meaning they are born with feathers, are well developed, and can walk and handle edible prey within a few hours of hatching. However, these precocial chicks are unable to thermoregulate, so they must be regularly brooded by an adult to keep their body temperature high enough during bouts of cold weather. The larger shorebirds become thermally independent within a week; the nestlings of the peeps can take two weeks to achieve this status. After two to five weeks, these little "runners" can fly. All this time they have been feeding themselves. Only in woodcocks and snipes do we see the adults feeding the nestlings, which, though precocial, do not learn to feed themselves as quickly as the other shorebirds.

Once the tundra-breeding shorebird nestlings are old enough to properly thermoregulate, they may be delivered into "crèches" (collections of young), which may combine young of different nests and even different species. Bar-tailed Godwits and Whimbrels do this, perhaps to aid in predator defense and to allow adults to aggressively forage away from their offspring to gain weight after a period of nesting starvation. After the crèches form, the adult shorebirds begin to depart for their southbound staging grounds—their first leg of autumn migration. Yes, the adults abandon their young entirely on the nesting ground, leaving them to find their way on their own to the wintering grounds, which may be as distant as southern Chile or Argentina. Crèche formation is presumably the social cue for the young to form a small society of parentless offspring, which must soon divine a way of getting to their wintering grounds safely. This is, indeed, not without its perils. Fewer than half of nestlings survive to the next summer.

I spent most of my time in the north bog, which had easier access because of an elevated road that led onto an abandoned gas drilling pad, which was set in the middle of the bogland. Walking this access road and wading out into the bog from the edge of the drilling pad provided lots of prime locations for shorebird-watching and photography. I made visits out there before and after each meal, seeing godwits on every single visit. And there was more. The gulls provided a continuous soundtrack. Large and vocal white birds were in the air nearly all the time.

The nesting Arctic Terns were of greater interest to me than the gulls. To me, the Arctic Tern is perhaps the most sublime of its lineage—the blood-red bill and feet, the graceful flight that highlights the long tail streamers and pointed wings, and the jet-black cap, separated from the dove-gray body by a slim white band below the eye. There is no more elegant North American tern than this one. The tern's ground nests were scattered about the boglands. Because the birds are mainly white, there is little subterfuge with the terns, in contrast to the shorebirds; when the adult tern sits on the nest, it is there for the world to see. Because of this, the adult terns must vigorously defend their nest. It is not uncommon for a nesting tern to strike a trespassing human on the top of the head with its bill.

Other shorebirds also nested in the bog. Diminutive Least Sandpipers carried out aerial display flights. And each morning the Wilson's Snipes winnowed high overhead. Short-billed Dowitchers scattered their nests in among those of the godwits. And Lesser Yellowlegs complained about my presence with never-ending scolding and head-bobbing.

Greater Yellowlegs summer in the boglands, along with the Lessers. Both nested among the godwits and dowitchers. Individuals foraged as singletons. The Lesser seemed the more skittish of the two, continually bowing its head and giving a shrill note of complaint when approached. In the bogs the Lesser seemed the more paranoid of the two on their breeding grounds. The yellowlegs are among the most familiar of North America's shorebirds, especially during spring and fall migration, when one finds them in small flocks, mainly Lessers with Lessers and Greaters with Greaters. Both exhibit the long, bright yellow legs in all seasons, making yellowlegs easy to pick out on a mudflat.

Bigger birds were there as well. Bald Eagles passed overhead and created a stir among the nesting birds. Northern Harriers sailed low over the bog in search of voles or lemmings. But most of the time it was gulls of various sizes in the air. Then, all of a sudden, a huge dark shape would catch my eye as a cow Moose appeared out from behind a hummock of spruces, lumbering across the bog on its tall skinny legs, inspiring me to look for an escape route or hiding place. Nobody wants to tangle with cow Moose, which are quite skittish in the spring because they are pregnant or with a newborn. On another afternoon a cow Moose suddenly appeared, tramping across the south bog, splashing through the water and dragging afterbirth.

Yearling Migrant Groups

Yearling shorebirds born that summer (known as "birds of the year") form flocks on the nesting ground and initiate their migratory movements after the adults have departed on their own migration. Not a lot is known about the southbound migration of naive juvenile shorebirds, but presumably these groupings stay together all the way to the wintering grounds. The ability of birds of the year to successfully migrate south and arrive safely at their wintering grounds is certainly one of the most remarkable and least understood phenomena in bird migration. Young tundra-born Pectoral Sandpipers have been found in the Pampas of Argentina in August, indicating these young birds travel long distances quickly and accurately. Apparently yearling Spotted Sandpipers maintain their association with flockmates even on the wintering ground.

When yearling groups launch their first long, southbound migration, is their chosen route hardwired or somehow "learned"? An experiment by Loonstra and colleagues with translocated yearling Black-tailed Godwits in northern Europe suggested evidence for learning or influence by association with godwits from another breeding site. Young hand-raised godwits from a Netherlands breeding stock were translocated to a Polish population that exhibited a differing timing of migration and a different wintering destination in Africa. The translocated birds mainly joined Polish flocks

The Yellowlegs
1. Lesser Yellowlegs ***Tringa flavipes***
Other names: Summer Yellow-legs, Small Cucu, Yellow-leg Tattler
2. Greater Yellowlegs ***Tringa melanoleuca***
Other names: Winter Turkey-back, Cucu, Horse Yellowlegs, Greater Tattler, Greater Tell-tale, Yelper, Tell-tale Snipe

These two shank sandpipers are among the most familiar and beloved of the North American Scolopacidae.

Appearance: The two yellowlegs species are nearly identical. The two are both slim, long-legged, active, and vocal. They are long-necked and small-headed. The Greater has a larger, thicker, and slightly upturned bill and a louder and longer series of call notes.

Range: The two species breed from Alaska across Canada to Newfoundland and Labrador, from tundra to taiga openings. The Greater has a slightly more southern and broader breeding distribution. The yellowlegs winter to the coasts and the southern sector of the Lower 48 as well as through Central and South America.

Habitat: The two species breed in muskeg and openings in boreal forest. Nonbreeding birds frequent shallow ponds in marshland.

Diet: Their diet is a mix of insects, crustaceans, and tiny fish. The birds are most commonly seen wading in shallow water of estuaries, ponds, or tidal creeks. They forage by delicately picking from the ground to chasing prey with their bill slicing through the water.

Behavior: The yellowlegs do an aerial display with accompanying vocalizations. Both are annoyingly loud scolders of human interlopers on the nesting grounds as well as on the foraging grounds.

Nesting: The nest, placed on the ground in a boreal opening, is a shallow depression lined with plant material. The female lays three or four buff eggs blotched with darker colors. Both adults incubate.

Conservation: The Lesser is included on the NABCI Watch List (with a declining population of 650,000). The Greater is not in decline and thus not Watch Listed (population 137,000).

of juveniles and traveled to the traditional Polish-bred destination in Africa. In spring, the translocated birds returned to Poland, not to the Netherlands. This experiment supports the notion that cues and information provided from conspecifics may be important influences on the initial autumnal migrations of juvenile godwits.

In the spring/summer of 2021, Senner and his team located 14 nests, captured 28 adult godwits, retrieved 12 geolocators from returning adults, and color-tagged the chicks on most of the located nests. Senner has been working on Hudsonian Godwits for more than a decade, and his intention is to keep working at Beluga until he solves the various riddles of the life history of this most remarkable species.

My visit to Beluga was an excellent counterpart to my 2019 stay in Churchill. There is nothing quite like standing in a wet bog where Hudsonian Godwits are working hard to produce an abundance of young godwits eager to carry on the lineage. My experience in the fen at Churchill featured territory-establishment and display by the godwits. The climate there was still winter-like, with freezing temperatures and iced-over puddles. It was too early for any arthropod abundance. Timing of their arrival is closely calibrated to the last of the extreme climatic conditions. The "early birds" win the race.

Spring was more advanced during my visit to Beluga, and nest-building was underway. The noisy display season I had seen at Churchill was over, and the nest-sitting season was in place in Beluga. The cycle of godwit reproduction was in full display. The main take-home point was the tightly regimented breeding schedule of the birds, driven by the demands of season, climate, and environment. ✵

8 Sub-Arctic and Arctic Outposts

The habitats chosen by spring shorebirds are extremely various. Grassland and prairie is favored by some, shores and marshes attract others. The Surfbird, Wandering Tattler, and Bristle-thighed Curlew like high alpine situations.

—Peter Matthiessen, *The Wind Birds*

IN THE SPRING OF 2022, my journey again took me north in search of godwits and other shorebirds, from my home in Maryland, to the Yukon, Alaska, and the Northwest Territories. This trip took me to the third breeding grounds of the Hudsonian Godwit (the Mackenzie River Delta of the Northwest Territories) and the breeding grounds of the Bar-tailed Godwit, Whimbrel, and Bristle-thighed Curlew (western Alaska)—four members of the Magnificent Seven. It also took me to the shores of the Arctic Ocean and the northernmost point on mainland North America. This was an adventure I had dreamed about for a decade, but a number of questions dogged me regarding driving solo and tent camping in the Far North. How does one avoid running out of gas when filling stations are few and far between? How does one store food to avoid attracting bears to the car or the campsite? What about hiking solo in Grizzly Bear country? And what if the car breaks down near the Arctic Circle, far from a repair shop? Worries about these unknowns kept me awake nights in advance of my departure. Perhaps that's what made this trip an adventure.

Opening the Google Maps app on my iPhone, I keyed in my location as "Bethesda, Maryland," and my destination as "Tuktoyaktuk, Northwest Territories." After a slight delay, the driving route popped up on the small screen. It showed a nearly straight northwestward route to the Yukon, then it bent to the north to the Arctic Ocean in the Northwest Territories: 4,434 miles and a drive of 76 hours. The app made it look so simple and straightforward! And it gave me hope it could be done. Just a matter of loading up the Xterra and turning on Google Maps.

COVID-19 zapped my travel plans in spring 2020 and spring 2021 because Canada closed its borders to miscellaneous travelers. So this twice-delayed field trip finally commenced at 6:03 a.m. on May 22, 2022—my biggest trip of the godwit adventure. A southerly route west bypassed Chicago's unpleasant traffic. Day One: 676 miles of driving got me to Urbana, Illinois (passing through Maryland, Pennsylvania, West Virginia, Ohio, and Indiana). Locust trees were blooming in profusion on the slopes above the Illinois River. My best bird of the day was a Red-headed Woodpecker (in Indiana). Iconic road signs for the day included "Jesus Is Alive" (Indiana) and "Guns Save Lives" (Illinois). In the evening, I took a stroll in the countryside outside of Urbana and listened to Dickcissels, Chipping Sparrows, Song Sparrows, and House Wrens. No shorebirds.

Day Two's travels commenced at 5:45 a.m., and encompassed 743 miles, passing through Illinois, Wisconsin, and Minnesota, and crossing into North Dakota, ending up in Fargo, as a thunderstorm filled the sky with dark purple clouds and flashes of lightning. My day's route (through Bloomington, Indiana; Rockford, Illinois; and Madison, Wisconsin) avoided the worst of the Chicago morning traffic. That worked, but I failed to anticipate the Minneapolis evening rush (you can't win). The day's highlights included the tall limestone stacks of the Wisconsin Dells and more masses of flowering locust trees. And also lots of roadkill: White-tailed Deer, Raccoons, and Striped Skunks. I dined in a dreary Mexican restaurant and stayed in a motel just across the parking lot from the sad restaurant. There was noise from people coming and going all night long.

From Fargo, I departed predawn for Moose Jaw, Saskatchewan, driving 534 miles northwest through Jamestown, Minot, and Portal (North Dakota), and Estevan (Saskatchewan). In Berea, North Dakota, a flock of eight Hudsonian Godwits greeted me from a wet field. The flock included several glossy adult males in breeding plumage, foraging in standing water. I passed through Velva, North Dakota, home of Dot's Pretzels. I did not stop to see Dot or the pretzel factory, but I appreciated the idea of these very tasty pretzels being manufactured in a place named Velva.

Now well into the prairie potholes, I passed lots of small ponds studded with pairs of Western Grebes. Swainson's Hawks hunted over the rich black-earth plowed fields, and in the towns Collared Turtle-Doves perched uncertainly on telephone lines. The drive crossed flat and wide-open spaces and huge skies with puffy clouds. Horned Larks sang and male Yellow-headed Blackbirds performed their noisy aerial display, rising from the tall reeds and then parachuting slowly downward. In the high prairie of Saskatchewan, Marbled Godwits foraged in a wetland and Upland Sandpipers hunted prey in some short-grass prairie.

It was 5:30 p.m., sunny, and 70°F when I arrived in the Wakamow Valley campground in Moose Jaw. Three days down and five to go to get to Fairbanks, Alaska, my first planned destination. The night was pleasant, but like so many campgrounds, this one was a stone's throw from some rail tracks, and freight trains rumbled by throughout the night, awakening me repeatedly. At dawn it was 43°F.

A juvenile Baird's Sandpiper in a South American wetland.

On May 25 I traveled to Alberta—a new Canadian province for me. This 573-mile drive took me from Moose Jaw, Saskatchewan, through Saskatoon and into Alberta, through Vermilion and Edmonton, across the main expanses of Canadian prairie, and into the first of the northland's vast boreal conifer forests, ending in Whitecourt, Alberta. Near Saskatoon a wetland hosted more than 150 Red-necked Phalaropes. These waters also offered up a pair of Cinnamon Teal and a Baird's Sandpiper. Some Pronghorn parties grazed out on the prairie. Alberta produced my first magpies of the trip. On this evening, driving through Edmonton, lots of Oilers flags were flying from passing cars because it was the night of a National Hockey League playoff game.

The Whitecourt Lions Campground provided me with a tidy tenting campsite, nestled in a stand of White Spruce and Jack Pine, right next to a boggy wetland. Whitecourt is apparently a timber and oil town, and I slept indifferently because the night was filled with noisy trucks on the move. The morning skies were clear, and at dawn it was 40°F. An early morning bird walk along the edge of the spruce bog generated a short list: White-throated Sparrow, Clay-colored Sparrow, Dark-eyed Junco, Yellow-rumped Warbler, and of course American Robin.

Day five took me another 521 miles, from Whitecourt, Alberta, to Andy Bailey Regional Park, just south of Fort Nelson, British Columbia. The route passed through Grande Prairie and Beaverlodge in Alberta and Dawson Creek, Pink Mountain, and Prophet River in British Columbia; I saw the last patches of prairie and entered northwestern forest lands thick with aspen and spruce. In Dawson Creek I picked up the Alaska Highway. Logging trucks pulling triple trailers of conifer logs headed southward on the highway; this is big-time timber country. Snowcapped peaks rose in the distance—the northern Rockies. Roadside birds included Common Raven, American Robin, Red-tailed Hawk, American Kestrel, and Red-winged Blackbird. Some Mule Deer and a Black Bear foraged by the roadside.

My evening's campground overlooked Andy Bailey Lake, which offered up various goodies: American Bittern, Bonaparte's Gull, Ring-necked Duck, American Goldeneye, Common Loon, Greater and Lesser Yellowlegs, Solitary and Spotted Sandpiper, and American Beaver. The surrounding spruce forest rang with voices of Ruby-crowned Kinglets, Ovenbirds, Yellow-rumped Warblers, Chipping Sparrows, and Common Ravens. The aspens were still entirely bare of leaves. Shady spots were marked by remnants of the winter's roadside snowpack. Five days of driving had gotten me to the north country. I was getting excited about my new surroundings but was not prepared for what was to come.

Day Six was one of the highlights of this trip; I traveled 475 miles across northern British Columbia and into southern Yukon.

This was larger-than-life country featuring snowcapped mountains in all directions, gorgeous glacially carved valleys carpeted with spruce forest, and big game everywhere. The Alaska Highway took me through Summit Lake, Toad River, Muncho Lake, Liard River, Coal River, Morley River, and Teslin Lake. This route passed through several provincial parks and along some rough unpaved road. Here was a big patch of wilderness that dominated the northwestern quadrant of North America. The scale of everything was overwhelming. It made this drive arguably the most majestic of my travels.

Traveling on my own, in my own car, completely self-contained and in charge of my daily doings, I thought back to the late 1950s and early 1960s when all my travels were dependent on elders (a grandmother, a mother, a father). Back then, when I glimpsed something interesting out the window, the adults invariably judged we were unable to stop the car to take a better look. I was never able to take it all in. Never able to get out of the car and take a picture or collect a sample from a roadcut. Now I could do all those things. What a happy evolution! The joy of traveling solo is the absolute freedom of movement—and freedom to stop and examine a point of interest more closely.

This was the ideal day for being on my own and taking it all in. Between Steamboat Mountain and Coal River, Mother Nature rose to the fore. Much of the route was through parkland, and the wildlife seemed to realize this. The road wound through mountain country, and I rarely saw another car. A Dusky Grouse stood in the middle of the road. A scattering of individual Black Bears foraged on emergent vegetation along the roadside. Each ignored me as I stopped the car and zoomed my lens in on it. Four small groups of Woodland Caribou ignored my passing car. I navigated gingerly through several groups of road-hugging Wood Bison, seeing old bulls, females, and bright rusty-brown calves as they worked the grassy verges, casually moving about with impunity. I saw a few Mule Deer. This is about as far north as the species ranges in North America. Two groups of Dall

Sheep fed at the roadside, and when I stopped to photograph them, they simply eyed me impassively. Most surprising, perhaps, were the porcupines; I spied eight of them, out and about and clearly prospering in this wooded boreal habitat. Muncho Lake, British Columbia, was mainly iced over, but a small opening of blue-green water held a pair of Barrow's Goldeneyes; in all my shorebird travels, this was my only encounter with this uncommon and retiring duck species. I pulled the car over and snapped a few photographs of this beautiful twosome in the icy green water.

Coal River Lodge, still in British Columbia, offered gasoline (I stopped whenever gasoline was being offered), and I took a break to dine on a tasty bison burger (perhaps recovered roadkill?). Tree Swallows flitted about nest boxes put up by the lodge. In Watson Lake, Yukon, at the end of the day, a repair shop (Rugged Repairs) was open, and the proprietor, Norm Leclerc, was willing to give my car a quick once over. He plugged the computer into the dashboard and checked the tire pressure. While Norm worked on the car, I watched Violet-green Swallows come and go as they constructed their nests in the workshop's eaves.

Norm reported that the tires were seriously underinflated and the computer had highlighted several issues, which he corrected. Ninety minutes later, I drove away, feeling considerably more confident that my car would not be breaking down on the Alaska Highway. Norm had fixed half a dozen problems that were either overlooked or were created by my over-priced Bethesda automotive repair shop. Right away, I got four more miles to the gallon of gas—a valuable benefit in this land of distant gas stations. I drove on another 150 miles and then set up camp beside the 40-mile-long sliver of Teslin Lake in the Yukon.

The lakeside campsite was bracketed by tall piles of snow, and the lake itself was icebound. The aspens were bare. The campground was not terribly birdy—just Common Ravens, Yellow-rumped Warblers, Orange-crowned Warblers, and Black-capped Chickadees.

The tent was cold all night, and the temperature hovered around freezing as the icy lake chilled the landscape.

On May 28 the road led northwestward toward Whitehorse, capital of the Yukon, where I stopped for breakfast in a decrepit but friendly diner just outside of town. Breakfast was excellent. This day the route followed glacially carved valleys bracketed by snowy mountain ranges that trended north-northwest to south-southeast. A drive of 484 miles took me from Teslin to Tok, Alaska, passing just east of the massive and snow-bedecked Saint Elias Mountains. I checked off a bunch of new destinations: Haines Junction, Destruction Bay, Burwash Landing, Quill Creek, Koidern, Snag Junction, and Beaver Creek in the Yukon and Northway and Tok in Alaska. The scenery was stunning and vast. Wildlife highlights were few. Aside from a single Black Bear, I saw a single Snowshoe Hare, as well as a Trumpeter Swan sitting on a nest of mounded vegetation in Tetlin National Wildlife Refuge, Alaska. Near Champagne, Yukon, two bicyclers welcomed a conversation on the lonely road. They were out for the long haul. They started their trip in Prudhoe Bay, at the top the Dalton Highway, on the Arctic Ocean of Alaska. Their destination was Argentina. These are the sorts of people you meet on the Alaska Highway. But during most of the day's drive the road was empty of humans and vehicles. It was just me on the open road, driving under the blue vault of the big sky.

About 35 miles east of Beaver Creek, a dark form lay motionless on the right roadside. I knew immediately that it was a road-killed Wolverine. I stopped to inspect this rarity, which some nasty driver probably struck on purpose (yes, there are some drivers who do that sort of thing). I photographed the carcass and hoped I would get to see a living one in the days to come. Being where Wolverine were roadkill meant I was in the big wilderness. In every direction were rock-strewn mountains with snowy tops.

Crossing into Alaska at Beaver Creek felt like a substantial accomplishment. Seven days of driving from Maryland to Alaska!

I set up camp that evening at the Tok River campground, in a nice stand of spruce, just outside of the hamlet of Tok. Alaska felt different—more like home. Red Squirrels played about in the trees by my tent. Boreal Chickadees, Hermit and Swainson's Thrushes, Canada Jays, Yellow-rumped Warblers, Dark-eyed Juncos, and American Robins passed by my tent. These are core boreal bird species. The only people I saw at the campground were a local family fishing for Burbot, an unfamiliar creature to me. They caught none.

Backwoods Alaska is distinctive. Lots of rural roadside homes that combine rustic habitation, repair shops, and yard-stacked solid waste storage—ah, the backcountry life. The first two Moose of the trip showed themselves on Route 2, on the drive to Fairbanks. But overall, Alaska highway driving was relatively wildlife-free; I suppose widespread rural subsistence hunting keeps most creatures away from the roadsides.

I spent the Memorial Day long weekend with friends Julie Hagelin and Peter Elsner in Fairbanks, and then I headed on to Anchorage. The long drive down Route 3 took me by Cantwell and Denali National Park. With the sky cloud-free, I was treated to various vistas of the great white mass of Denali. At one point my rearview mirror showed Denali, with tall spruce neatly framing the giant summit. Through 11 days of travel, there had been only brief minutes of rain. The weather had been ideal for cross-country travel.

Back once again in Anchorage, and camping at Eagle Creek, just outside of town, I prepared for my first adventure of this year's journey—a flying side trip to Nome, a springtime birder's paradise promising chances to encounter a Bar-tailed Godwit and Bristle-thighed Curlew.

Highlights of the brief Anchorage stopover included a single Hudsonian Godwit out on the flats of Cook Inlet, a close encounter with a confiding Black Bear on the Arctic Valley road, and a congregation of five bull Moose in a field at Point Campbell Park, not far from a cow Moose with three babies, down by the airport. That's nine Moose in one morning—all within the confines of the city of Anchorage. It seemed the Alaska Moose had a good sense of where big game hunting was prohibited. After dropping my car off at the Nissan dealer for a top-to-bottom checkup, I flew to Nome, with hopes of novel shorebird encounters.

The 90-minute flight from Anchorage to Nome took me over the snowy peaks of the Alaska Range and northwest across rough country that was a mix of tundra, lakes, ponds, and conifer forest, with no sign of human habitation. We touched down at the airport with thick fog blanketing the Bering Sea and the noon temperature at 39°F. Nome is on the south shore of the Seward Peninsula in westernmost Alaska, sitting atop the Bering Land Bridge. Siberia lies about 175 miles due west. Saint Lawrence Island lies about 120 miles to the southwest. The human settlements on the Seward Peninsula are mainly coastal. The interior is largely undeveloped today, though there was an era of alluvial gold-mining back in the twentieth century that ended in the 1960s. Today, the main economic drivers are tourism and subsistence activities carried out by Indigenous communities. The rivers support small salmon runs. In late May and early June, birders from around the world descend on Nome to revel in the arrival of spring and the abundance of migratory landbirds and waterbirds that nest there.

Best known as home of the finish line for the Iditarod Trail Sled Dog Race, Nome is a small town, unprepossessing when the snows melt in spring. Early spring in snow country is grimy, and in Nome the grime was on full display. When I arrived, the several hotels were filled with tour groups of birders from the United States, the United Kingdom, and Europe. All the rental cars and vans were deployed to birders who relentlessly drove the 150 miles of roads in search of Alaskan bird species to add to their life lists. The road system is simple. Three roads spread out from Nome: one northwest to Teller, one north to Taylor (the Kougarok Road), and one northeast to Council. Virtually every mile of this mainly graveled road system is good birding. So for me it was drive, stop, bird, repeat—for five days. The landscape was open hilly tundra, with the only woody vegetation mainly being the willow thickets that filled the stream valleys.

Nome drew me mainly for two of the Magnificent Seven: the Bar-tailed Godwit and Bristle-thighed Curlew. These two Pacific super-migrators breed in the tundra landscape of the Seward Peninsula; the Bar-tailed also breeds in Siberia. The Bristle-thighed breeds exclusively at a couple of upland locations: one on the Seward Peninsula and the other on the next peninsula south of the Seward (home of the Yukon River). The Bar-tailed is the migrant champion of shorebirds. The Bristle-thighed is one of the rarest and least known of our shorebirds. The chance of spending time with these two in their nesting habitat was a special privilege.

In Nome it was necessary for me to adapt to a different living situation. There is a single campground (Salmon Lake) outside of Nome, but getting all my camping gear there by plane would have been expensive. And since Grizzly Bears are the common ursid in the countryside, most people are not eager to tent camp there. So in Nome I slept in a hotel, ate in a restaurant, and drove a rental car.

I did a lot of driving and birding from the roadside. Nome at the end of May was very birdy and mostly snow-free. The Bering Sea was ice-free, and the only remaining ice was that which had rafted up on the beach during the winter. The common roadside birds were in song and in display. Streamside willow thickets hosted singing Gray-cheeked Thrushes, Northern Waterthrushes, and American Tree Sparrows. Handsomely plumaged male Lapland Longspurs were everywhere, doing display flights and singing to beat the band. And Semipalmated Sandpipers were stationed in the roadside lowland tundra, doing their helicoptering display and making their weird trilling that sounds like some malfunctioning electronic device. The Longspur males were more beautiful, but the Semipalmated Sandpipers doing their over-tundra nuptial display won the day.

My priority target bird on the Seward Peninsula was the Bar-tailed Godwit, the third godwit species breeding in North America, counterpart to the Marbled and the Hudsonian. The species breeds in the Arctic from northern Sweden and Finland east to western and northern Alaska, wintering to South Africa, South Asia, Australia, and New Zealand. In Alaska, the species nests in hilly tundra around Nome. I drove the roads through the coastal tundra in search of godwits on territory but found not a one. On several different days I walked roadside tundra tracks in search of Bar-tails but only heard of other birding parties finding them on territory. They were there, but this big rusty-breasted shorebird was elusive. I did manage to see other good birds nesting on the tundra, including the Pacific Golden-Plover and Whimbrel.

And another iconic Nome breeder provided some solace to my disappointment with the Bar-tailed Godwit. It was the Long-tailed Jaeger. Worldwide, this is the rarest of the three jaegers, and I had seen the species but once, at sea, off the coast of North Carolina. That was an all-dark, short-tailed young bird. To my mind, the breeding adult is the most beautiful of the world's seabirds. Its long and gracefully pointed tail, slim pointed wings, and bold and simple plumage pattern render this bird a marvel when in flight.

The world's three jaegers, gull-like predators that nest in the Arctic and winter out over the world's seas far from land, are not familiar to many birders. Nome is home to all three, but the Long-tailed is the most common. I encountered individuals several times a day in both the lowlands and the uplands. I even found an active jaeger nest not far from the Teller Road. The single large olive-green egg rested in a slight depression in the tundra vegetation, there for everyone to see. The nest was obvious because the adult was sitting upon it, the bird's head and back well above the tundra vegetation.

Every time this smallest of the jaegers passed overhead, I leaped from the car with my long lens, snapping the bird in flight, set against the deep blue sky.

But back to the Bar-tail. My Alaska bird-finding guidebook told me that my target godwit spends a lot of time foraging and loafing in the flats at the mouth of Nome Creek, just on the edge of town. I had been told the godwits were regularly visible from the Council Road where it bridged the creek. In the morning I stopped on the bridge to look for godwits and indeed there they were. But parking was difficult, and viewing conditions were challenging with the backlighting and the distance as they waded in the shallows, probing the mud. I never got a great photograph of this gorgeous godwit, and missing this species on its tundra breeding ground was a disappointment.

Although the bulk of my time was spent hunting for shorebirds nesting on the tundra, I was entertained by other birds that I had little experience with. On my second day in Nome, I spoke with a visiting group of excellent young birders (Nathan McGowan, Dan Maxwell, Steve Tucker, and John Garrett). We were all dining in Milano's Pizzeria—"the best Italian, pizza, Korean, Japanese, sushi, burger joint in town." They told me that the first bridge across the Salmon River on the Council Road hosted an active Gyrfalcon nest.

The Gyr is the largest of the world's 64 species of falcons and the most northerly breeding of the lineage. It is a bird that is only rarely seen in the Lower 48. The next morning I drove out to the bridge, parked the car, and climbed down under the bridge to look for the nest. There atop a concrete pillar in the shade was a messy stick nest topped with three fuzzy white nestlings. I sat on the riverbank and waited. Before long in sailed a big adult. It circled over the bridge, eyeing me suspiciously. I headed back to the car and climbed in. The falcon dipped down to the nest, and I returned to photograph the adult with the young. A second adult approached overhead. It was powerfully built, much more massive than a Peregrine. Its wings were pointed but broad-based; its face was marked with a dark moustachial stripe and something of a dark hood. The circling bird gave a harsh rising series of slow shrieks, expressing its dismay. I retreated from the scene with my treasured pictures. The two adults were of the gray morph, the common plumage color in North America—plain gray above and roughly dark streaked below, with heavily patterned underwings.

On my third morning in Nome, I rose before dawn and drove the Kougarok Road into the interior uplands of the Seward Peninsula. I was headed along 69 miles of rough gravel road inland to Coffee Dome, known to be the most reliable site for the ever-elusive Bristle-thighed Curlew. Between mileposts 35 and 50 scores of Snowshoe Hares appeared in the road and by the roadside. Then, at mile post 51, a Canadian Lynx posed in a willow thicket atop a roadside hillock, intently watching my car. Rolling down the window, I started shooting pictures of this big tassel-eared cat. I had seen Bobcats before but never a lynx. Seeing the many Snowshoe Hares followed by the lynx—a notorious hare-hunter—was particularly pleasing, given how much has been written in the scientific literature about the prey-and-predator relationship linking these two species.

This lynx, hiding in the thicket, stared at me for several minutes, then raised up and showed its head and shoulders. It then leisurely jumped down off the hill and sauntered across the gravel road and into the bushes on the far side. After the cat disappeared, I just sat in the car and reviewed the 36 images on my memory card. The pale olive eyes! The tall black tufts atop each ear! The oversized feet! The black-tipped bobbed tail!

Arriving at the trail head for Coffee Dome in the very early morning, I hiked up onto the tundra ridge in search of Bristle-thighed Curlews. For quite a while I was the only person on the mountain. I walked the tundra for three hours, not seeing another human until I was heading downhill. Instead of the Bristle-thighed, I found territorial individuals of the nearly identical Whimbrel, identifiable because of its distinct voice. I also spent quality time with two breeding pairs of American Golden-Plovers—featuring the glossy black belly, throat, and face, the white trim, and the golden-flecked back and wings. A male carried out a series of display flights overhead, all the while calling its sad upslurred whistles. On the ground, this glorious bird pranced nervously in the low tundra, bracketed by the rich pinky purple of lousewort flowers. Recall that I had encountered a flock 249 of these beautiful plovers in a plowed dirt field in South Dakota. Here was where those birds had been heading—upland tundra expanses such as Coffee Dome.

Driving back south toward Nome, I was disappointed that I had seen no Bristle-thighed Curlews. Yes, I had heard the distinctive flight call of the species, but just once and from a distance. My disappointment was mollified when I spied a bull Muskox and then noticed a small herd of females and young of this strange Arctic-dwelling bovid out in a tundra valley below the road. Muskox were exterminated from Alaska by overhunting in the nineteenth century. These unusual tundra foragers are found today on the Seward Peninsula because of a reintroduction effort by the US Fish & Wildlife Service in the latter half of the twentieth century.

Seeing these dark-furred creatures, dark dots on the vast expanse of tundra, reminded me that these open lands once were populated by Woolly Mammoths, Woolly Rhinoceroses, and other mammalian giants until they were hunted to extinction by humans. Did it matter that the Muskoxen I was seeing were the product of a reintroduction? Heck no! I was just happy that they were on the landscape, enriching it. Moreover, I am hoping molecular biologists will clone the Woolly Mammoth and Woolly Rhino before too long. What creatures to behold on the tundra landscape! I would pay for that.

Two days after my first jaunt to Coffee Dome, I decided to try again. I rose at three in the morning to ensure I was the first birder on the mountain; I pulled my car up to the high pass after two and a half hours negotiating the winding gravel road. Departing Nome, the temperature was 46°F. Up on the interior ridge, it was 30°F.

Bristle-thighed Curlew *Numenius tahitiensis*

For many decades, the Bristle-thighed Curlew remained one of the mystery birds of our continent. Its nest was finally discovered in the lower Yukon River uplands of western Alaska in 1948 by David Allen and Henry Kyllingstad.

Appearance: This species looks like the Whimbrel, which can be found sharing the Bristle-thighed's restricted breeding habitat. The Bristle-thighed can be distinguished by its cinnamon rump and the stiff bristles that protrude where the upper legs meet the body.

Range: The breeding population is confined to two small patches of interior uplands in western Alaska—one on the Seward Peninsula, the other near the mouth of the Yukon River. After staging in the Yukon-Kuskokwim Delta in mid-summer, the adults migrate south across the Pacific to various islands in the equatorial Pacific.

Habitat: The species nests on tundra hills. Wintering birds inhabit tropical beaches, reefs, and islets of the South Pacific.

Diet: The bird's diet while nesting on the tundra is mainly arthropods and berries. Postnesting, adults stage in the coastal zone of the Yukon-Kuskokwim Delta to forage voraciously on a range of ericaceous berries and invertebrates to build up reserves of fat for the long Pacific flight. Wintering birds forage on crustaceans (especially ghost crabs), birds' eggs, mollusks, and tiny fish.

Behavior: The monogamous male carries out a vocal display flight over its large breeding territory. The flight song is a high musical series of slurred notes that rise and fall, quite distinct from that of the Whimbrel. This is the only shorebird to winter solely on oceanic islands and the only one to become flightless during molt.

Nesting: The Bristle-thighed's nest is a depression in the tundra vegetation, lined with various bits of plant material. The female lays four olive eggs with dark blotches. Both sexes incubate the eggs and spend some time caring for the young after hatching.

Conservation: The Bristle-thighed Curlew is the rarest North American shorebird (species population 10,000).

Again, stalking the frosty upland tundra, I was enchanted. I walked and hunted for two and a half hours but still saw no Bristle-thighed Curlews. But other birds entertained me. A bright, rusty-tinted Western Sandpiper perched on a tiny tundra hillock of variegated vegetation, declaring his breeding territory. It behaved like it was king of the hill. Solitary Long-tailed Jaegers sailed overhead. Whimbrels and American Golden-Plovers cried out, declaring their territories from overhead. Willow Ptarmigans, Savannah Sparrows, and Lapland Longspurs enlivened the walk.

As eight o'clock approached I spied a birding tour group near the western summit of the dome. From their behavior (bent over telescopes on tripods), I had an inkling of what they were looking at. It was a stiff 15-minute walk in ankle-breaking tundra to hustle up to the spot. The birders were now making their way in a straggling line back down to their van. Their leader (who turned out to be the famous British field ornithologist Craig Robson), still bent over his scope, pointed to a distant shorebird on the tundra. He generously said, "It's all yours!" I spent five minutes photographing this adult Bristle-thighed Curlew before it rose from the ground and circled, calling its diagnostic song phrase, a repeated and musical *tuuuu-whit-o-wheet* over and over as it headed eastward to parts unknown.

The winter range of the Bristle-thighed extends from west to east across the breadth of the tropical Pacific—from Micronesia and northern Melanesia to Easter Island in easternmost Polynesia. Data from tracked birds show the species moves southward on two distinct tracks, one to the southwest ending in Micronesia and the other directly south ending in Polynesia. Both these groups of southbound migrants carry out long overwater flights that can exceed 6,000 miles. It appears that most or all breeding birds stage in autumn in the Yukon-Kuskokwim Delta of western Alaska. That the southbound migrants, departing from a single staging site, diverge to follow disparate western and eastern destinations is mystifying. Is it possible that the birds of the two tiny and circumscribed breeding grounds fly to disparate wintering grounds on either side of the Pacific? Researchers assume the current breeding range of the species is relictual (a small remnant of what it was in the past); perhaps these side-by-side breeding ranges are remnants of two distinct populations that were formerly large and expansive—one from Siberia and one from Alaska, and perhaps the distinct wintering grounds reflect this history (in the manner that Siberian and Alaskan Bar-tailed Godwits winter on different sides of Australia).

I have read in reference books about the tundra's importance to shorebirds, plovers, ptarmigan, jaegers, and other birds as prime breeding habitat. After more than half a century of reading about avian tundra breeders, it was wonderful to have a chance to get onto that tundra and commune with these far-northern-breeding birds. While on the Seward Peninsula, I spent part of every single day walking the tundra—traipsing for hours across this seemingly featureless landscape that, in fact, displays plenty of variety and character. Most of the Seward Peninsula is tundra. Some of it is boggy and difficult to navigate (the trek up to the top of Coffee Dome is famous for being fiendishly difficult for older hikers). However, most tundra is easy to travel across, and it was simply splendid being out on the tundra, all alone, not another human in view.

Why do all those birds love nesting on the tundra? I think I found the answer during my long walks out there. Birds want a safe place to bring up their offspring. The tundra offers this in spades: vast open spaces! Tundra stretches from Arctic Scandinavia eastward to northern Alaska and onward across Canada to Hudson Bay and northern Labrador. These are huge expanses of undeveloped and open land with very long days in late spring and early summer. The long days provide an abundance of time for feeding nestlings or, in the case of precocial birds like curlews and godwits, for watching over the young runners while they forage for tundra invertebrates all day long.

But back to the wide-open spaces. The vast tundra landscape allows each nesting pair of curlews or godwits to spread out with minimal interference from pesky predators or conspecific competitors. Out on the tundra, here is what I found—lots of nothing. The density of nesting birds is astonishingly low. My long walks across the tundra turned up very little each morning or afternoon. The birds were few, and of course, they were also hiding from me, nearly invisible in the low vegetation. In fact, the tundra is an impoverished breeding habitat with few breeding birds per square mile. Walking for hours in the coastal tundra, I came upon not a single Bar-tailed Godwit. And their rarity is a saving grace for these breeding birds; this keeps populations of predators low.

Acknowledging the impoverished nature of the tundra, why do birders see lots of birds on the Seward Peninsula in late spring? Three reasons. First, the coastal wetlands are popular staging and foraging locations—distinct from the featureless tundra. Second,

An adult Sanderling in breeding plumage on its tundra nesting habitat.

spring sees lots of avian movement as migrants stream in. And third, the birders cover considerable distance by car to enrich their daily birding lists. If they had to do this on foot, it would be a different story. I saw lots of cool birds on the Seward Peninsula, but to do so I had to drive hundreds of miles.

On my last day in Nome, I drove down to Nome Creek to photograph the Bar-tailed Godwits. Indeed they were there, foraging in the mudflats. This was the reliable place to see these rare birds, not in the vastness of the tundra. The adjacent beach also welcomed flocks of terns, mainly Arctics but also a few Aleutian Terns. They were loafing there, allowing close approach. The Aleutian Tern's white forehead, black cap, and gray breast make it a very pretty tern (though it cannot compete with the Arctic). In flight the Aleutian Tern gives a high, rapid bubbling series of notes that descends in pitch. As with the Bar-tailed Godwit, one needs to travel to Alaska to have a chance to encounter this rarity in North America. Who knows where these Aleutian Terns will be nesting? It was just a special treat to be able to spend some time with them on the beach at the mouth of Nome Creek. That's a species I never expect to see again in my lifetime.

From Nome I returned to Anchorage, picked up my Xterra from the Nissan dealer, and bedded down at the Eagle River campground for the night in preparation for a flight the next day to Utqiagvik, Alaska, formerly known as Barrow. Birding in Anchorage the morning before the flight, I managed to see a single Hudsonian Godwit on a distant mudflat from parkland behind Ted Stevens Anchorage International Airport. My scheduled two-hour flight would take me to the northernmost point on mainland Alaska—the apotheosis of the High Arctic. Nome was intense. But Utqiagvik was extreme, even in early June.

The flight took me over an endless wilderness that transitioned to being ever more wintery as we lumbered northward under gloomy skies. After touching down at four o'clock in the afternoon at the Utqiagvik landing ground, right in the middle of the city, I walked to the rental car office, then drove to my downtown hotel. The air temperature was 28°F, there were snow drifts all around, and light snow was falling. We were right on the shore of the Chukchi Sea, whose shoreline was entirely iced over.

The plane that brought me here was loaded with birders from all over, many of whom had just done the Nome stop. I bumped into Clive Harris and Dave Powell, acquaintances from my local bird club in Montgomery County, Maryland. Finding them was a nice surprise. We agreed to coordinate our birding efforts over the next few days.

I was there for seven nights, and the days and nights tended to blur together, given the thick cloud that sat over us for much of the time, plus the sun's inability to dip far below the horizon that time of year. Rather than a detailed day-by-day set of reports, what follows is a general recollection of the highlights of early spring in Utqiagvik.

An adult female Red Phalarope in her bright breeding plumage.

Most of the landscape remained snow-covered, and the lakes and ponds were ice filled with small patches of open water. The town itself, being warmer than the surroundings, had more snow-free territory and thus tended to attract arriving bird migrants. Other substantial snow-free spots were roadbeds and roadside ditches. These were where we saw most of our birds. Many of the migrants had arrived before the landscape seemed ready to receive them, and this indeed made it easier for us birders to find the birds that were there. Moreover, Utqiagvik has only about 11 miles of usable roads, so our birding options were limited: the shore road northeastward toward Point Barrow, the Gas Well Road, and the gravel pit were the main venues, with ponds and lakes scattered about. We visited these over and over, looking for newly arrived migrants (new arrivals showed up every day). Another thing: at the top of the world there are not a lot of breeding bird species. My Utqiagvik list was 25 species versus my list of 69 species from Nome.

What of the shorebirds? I saw no godwits or curlews, which was a disappointment but not much of a surprise. Whimbrels and Bar-tailed Godwits breed in the area in small numbers, but they eluded me. Eight species of shorebirds did oblige, however. Gorgeously bright adult female Red Phalaropes twirled in tiny roadside puddles, making the best of a situation in which tundra breeding habitat was not yet free of snow. They were the most confiding of Utqiagvik's shorebirds. Red-necked Phalaropes were there in good numbers as well. A single breeding-plumaged Red Knot hung out in the middle of a gravel road, as if waiting for spring to arrive. Male Pectoral Sandpipers were already doing their strange flight displays along the Gas Well Road. Their expanded breast, like a large dewlap, was rather grotesque to my eye. The sounds they emitted were equally peculiar. The male Pectoral Sandpiper of Utqiagvik was a wholly different bird from the Pectoral Sandpiper I knew from the Lower 48.

Semipalmated Sandpipers and Western Sandpipers were scattered about the wintery habitat, as were clots of handsome-looking, black-breasted Dunlins, ready to rock and roll. The only other shorebird I encountered was the Red-necked Stint, a Siberian breeder that summers in small numbers in northwestern Alaska. In Utqiagvik at that time of year, it was too early for much reproductive activity. At this point the birds had arrived and hunkered down to wait for conditions that would permit breeding. This is that "hurry up and wait" phenomenon that happens in spring, when the males don't want to miss out on a chance to nab a territory and grab a mate.

What of other species? Ducks and geese form a major segment of the breeding avifauna in Utqiagvik. All four eider species show up in spring as the snow melts. The adorned males are a feast for the eyes. And where else on Earth can you see four eider species in a morning? Rarest is the Spectacled Eider, the big drake exhibiting its pale green crown and forehead. Steller's Eider is the smallest of the four, the males with golden flanks and harlequin upperparts. The drake King Eider is the most beautiful of the four, with its bright orange forehead. Common Eiders streamed overhead in large flocks in the early morning, heading to their scattered breeding grounds. Greater White-fronted Geese hung out in pairs in yards in downtown Utqiagvik, plucking at the brown grass. Pomarine and Parasitic Jaegers patrolled the snow-covered tundra. I watched a pair of Parasitics hunt down a Semipalmated Sandpiper and then methodically pull it to pieces—nasty hunters savoring a tasty tidbit. Snow Bunting was the commonplace town songbird. The black-and-white adult males sang lustily from the roof peaks of urban buildings and foraged in people's yards.

The occasional Arctic Fox scampered across the snow, looking rather drab in its gray spring coat. We encountered only one other mammal here—a Polar Bear. This grand beast had been drawn into town by some caribou meat and bone that was discarded out on the ice a few dozen yards off the town beach. The bear drew a crowd. An Iñupiat resident told me that there were dozens of bears out on Point Barrow, where the remnants of traditional Bowhead Whale kills had been left to scavengers (one needs a permit to venture out onto the Point, which is restricted Indigenous territory). Seeing a Polar Bear was an unexpected highlight, and it gave me a North American ursid hat trick on this trip (having seen many Black Bears on the drive north and a single young Grizzly on the Kougarok Road of the Seward Peninsula).

Back in Anchorage after my Utqiagvik trip, I picked up my own car from the airport parking lot and loaded it up with stuff stored at Dave Sonneborn's home. After reprovisioning at Carrs grocery, I headed northeast toward Canada's beckoning Yukon and Northwest Territories. I was now on my way to the Arctic Ocean of Canada, in search of Hudsonian Godwits on their North Slope tundra breeding territories. Northwesternmost Canada hosts a third breeding population of the Hudsonian, and I wanted to spend time with these birds on their breeding grounds, as I did in Churchill and Beluga.

Departing Anchorage in the evening, I drove up Route 1 along the Matanuska River to Glennallen, camping that night at the Dry Creek State Recreation Site. The drive with the lowering sun was spectacular, with the Chugach Mountains to the south. At one point the drop-off was more than a thousand feet into the gorge of the Matanuska—with no guard rail. The sun set on the Matanuska Glacier, which was ever-so-slowly pouring down out of a high valley into the Matanuska gorge. The driving route eventually broke through the mountains and entered a vast lowland plain of bogs and ponds of the Copper River. The forest of spruce where I parked my car at midnight disgorged swarms of eager mosquitoes—too many to allow me time to set up my tent. I slept in the car, the windows shut tight.

The next morning, in the freezing chill of dawn, the songs of Ruby-crowned Kinglets and Swainson's Thrushes awakened me. I breakfasted alfresco at Glennallen junction, enjoying an amazing vista of the massive snow-shrouded Mount Sanford (16,237 feet) of the Wrangell Mountains just to the east. After breakfast, the drive took me northeast, back to Tok, and then onto the Taylor Highway across the "Top of the World," through Chicken, Jack Wade, Boundary, and, at long last, Little Gold, Yukon, which marked the Canadian border. The road was rough and unpaved, slicing through hilly country much scarred by forest fire. There were few birds or other wildlife over this long and lonely drive; a Black-billed Magpie was the only bird that crossed the road. Once I was in the Yukon, a barren landscape of dun hills of upland tundra took over. This day's trek was, no doubt, the most disappointing stretch of road that I endured in the Far North for its lack of wildlife or scenic vistas.

At day's end I set up camp at the crowded Yukon River Campground, right smack on the banks of this big turbid river, with Dawson City on the far side of this flow of icy water. The Yukon drains the western flank of the Wrangell Mountains and curves north, northeast, and then southeast across the main bulk of Alaska. After joining with the Kuskokwim River to produce a vast lowland delta east-northeast of Anchorage, on the northern flanks of the Alaska Range, it dumps into the Bering Sea. This, the "Y-K Delta," is a famous staging ground for shorebirds in late summer.

The campground was at capacity, as Dawson City is a major tourist draw at this time of year. The campground's only avian highlight was a singing Townsend's Warbler—a widespread species of the mountain west that reaches its northernmost breeding range right about here. This golden, black, green, and white songbird was a beauty and new for the trip list. Recall that I last saw a wayward spring migrant of this species on South Padre Island, Texas.

After a night in the campground, I packed up my belongings and took the ferry across the big river to visit Dawson City to do laundry and shopping and to visit the local Yukon tourist bureau. The city, once powered by alluvial mining for gold, is now driven by summer's rush of wandering tourists who like to get off the beaten track—mainly folks in RVs, married couples in their seventies. The town was cute and tidy, and 100 percent oriented to the tourist trade.

I started my morning at the town RV park, where I washed a batch of filthy clothes. Then I visited all the shops to pick up odds and ends. Since I was heading into big-time Grizzly country, I purchased some "bear-bangers"—mini-projectiles that explode in the air to scare off approaching bears or Moose. Bear-bangers are a nice accompaniment to bear spray, an essential tool, which was already stashed in my car, along with Bear Shock—a portable bear-proof electric fence to erect around my tent. One cannot be too safe in Grizzly country!

In the late morning, after getting some great intelligence from the tourist bureau on birding the Dempster Highway, I headed east on the North Klondike Highway, following the course of the Klondike River (a tributary of the Yukon). The river is paralleled by high linear mounds of large, rounded river stones—the product of a bout of massive stream dredging to extract alluvial gold. I have seen similar stone piles in the Bulolo Valley of Papua New Guinea, where similar huge dredges sifted the river gravels to extract tiny grains of gold. The Klondike gold rush kicked off in August 1896 and lasted three years. A hundred thousand eager migrants came in search of fortune. A similar gold rush took prospectors (including a young Errol Flynn) to Papua New Guinea in the 1920s and '30s. The lure of gold is a powerful incentive to impoverished dreamers the world round.

A few miles east of Dawson City I came upon the entrance to the Dempster Highway, a 478-mile-long gravel road that would take me north beyond the Arctic Circle and toward my ultimate destination—the shores of the Beaufort Sea, an embayment of the Arctic Ocean on the eastern verge of the vast Mackenzie Delta—home to breeding Hudsonian Godwits. The Dempster transects the Yukon and the Northwest Territories and ends at Inuvik, the principal city of the northern Northwest Territories. A newly constructed road extension would then lead me from Inuvik northeast to Tuktoyaktuk, set on the shore of the Arctic Ocean, an additional 94 miles of driving on gravel roadbed through the Canadian tundra. This is the one place on the continent where one can drive to the northernmost Arctic coast of North America. The Dalton Highway, north of Fairbanks, Alaska, is similar but without a permit one can drive only to Deadhorse, a few miles short of the Arctic Ocean.

The drive up the Dempster Highway to Tuktoyaktuk had been on my mind for more than two years. Aside from the service stations at each end of this stretch of gravel highway, there were just four gasoline stations on the route: Eagle Plains, in the Yukon, and Fort McPherson, Tsiigehtchic, and Inuvik, in the Northwest Territories. The first stretch, at 230 miles, from North Klondike Highway turnoff to Eagle Plains, happened to be the longest. A tankful of gas would theoretically get me 290 miles in my gas-guzzling Xterra. But would poor driving conditions on the gravel road substantially reduce my gas mileage? At the entrance to the Dempster, I topped up my gas tank and began the drive north, following the valley of the North Klondike River. Not many places in North America are as distant or as isolated as this route north. Just to add to my anxiety, the tourist bureau adviser in Dawson City said that the Peel River Ferry at Fort McPherson was not operating because of ice choking the river. This ferry only operates after the spring breakup of the ice, and this was a late spring, apparently. When would the ferry open?

After 45 miles a signboard announced Tombstone Territorial Park, with its excellent interpretive center and affiliated campground, all tucked into the bottom of a broad mountain valley with peaks rising in every direction. The valley was vegetated in a mix of shrubland, tundra, and open stands of spruce. Dark rocky scree and snowdrifts patterned the higher ridges and summits. The river draining the mountains to the west (which includes Tombstone summit) ran down a glacially smoothed valley, and it was bracketed by remnant icepack that glowed bright white against the greens and browns of the low vegetation. This was my first stopping point on the Dempster, as it offered one of the few campgrounds along this lonely route.

I set up camp, chatted with the staff of the interpretive center about birds, mammals, and best hikes, and started exploring the heart of Tombstone Territorial Park (so named because the Yukon is a territory of Canada, not a province). This spot, in the mid-afternoon sunshine, with azure-blue skies, was a heavenly Canadian destination. It was nothing like I had imagined over the two years I had waited to visit—much more rugged and glacial and green and welcoming. I had expected stark tundra of brown vegetated flats. Instead what I found was one of the prettiest wild places in North America—a legitimate competitor with Yosemite, the Tetons, and Denali.

The campground was filled with travelers headed north in recreational vehicles, on motorcycles, and in station wagons. Many were Europeans on a North American adventure—not gold prospectors, but senior citizens prospecting for unknown places and special experiences. Most of these travelers were retirees, and all of us were looking for something new and unusual. Hence the drive up this dead-end gravel route in northernmost Canada. What golden lode would we find under the mythical rainbow at the road's end?

I spent several nights at the park and invested much of my time walking the trails that radiated out from the campground. I also wandered on bicycle (battling gravel and long grades) and by car (headed out in search of roadside wildlife). My local hikes took me through a mix of streamside spruce forest and shrubland. I hiked with my bear spray in a hip holster. The only shorebird I encountered was the Spotted Sandpiper—about as far north as this species ranges; the bird teetered atop a beaver lodge in a pond just downstream from the interpretive center. Most of the birds I encountered were familiar ones from home—Common Ravens; Yellow, Yellow-rumped, and Orange-crowned Warblers; Fox and White-crowned Sparrows; and Gray-cheeked Thrushes. The only unfamiliar species was Hammond's Flycatcher. I hunted for a Northern Hawk Owl that had nested near the interpretive center, but no luck. I also did

Wandering Tattler *Tringa incana*
Other names: Heteroscelis incanus

Appearance: The Wandering Tattler is a handsome two-toned sandpiper with fine gray zigzag barring on its underparts.

Range: The species summers in the Yukon, Alaska, and eastern Siberia, and it winters on the West Coast and southward through the Pacific to Hawaii and Australia. Many tattlers migrate across the Pacific to Australasia and the Pacific Islands. Some winter along the South American coasts.

Habitat: The Wandering Tattler breeds along tumbling rocky upland streams in the Far North, but it is most often seen on rocky or pebbly Pacific shores during the nonbreeding season. Its nest and eggs were not discovered until 1923. The bird can be found in association with Black Turnstones and Surfbirds in winter along the Pacific coast of the Lower 48, but the tattler is a more active and energetic forager. Tattlers are rarely seen in flocks.

Diet: The tattler's summer diet is mainly a mix of aquatic arthropods and small mollusks taken on streamside gravel banks. Its nonbreeding diet is mostly crustaceans and marine worms and mollusks taken on algae-coated rocks or jetties.

Behavior: The species is monogamous. The male displays over his breeding territory with a towering flight and attendant vocalizations.

Nesting: The nest, set in gravel along a mountain stream or in adjacent tundra, is a depression often lined with plant material. The female lays four olive eggs, heavily blotched with brown. Both parents brood and care for young nestlings, but then one parent departs, leaving the other in charge. The young feed themselves, often in close association with the parent.

Conservation: The species is classified on the IUCN Red List as Least Concern, with a stable but tiny global population that is apparently no more than 25,000.

evening playback for Boreal Owls several times in the dusk of "night" but got no response. I saw no mammals beside Red Squirrels. I photographed an unfamiliar butterfly—a Margined Mustard White.

Two Arctic-breeding shorebird species nest in the park, Wandering Tattler and Surfbird. The tattler frequents rocky mountain torrents, habitat where one also finds the American Dipper. A few miles north of the campground the highway follows just such a stream, and I spent several hours on two different occasions hunting for the tattler. This was a bird I had known as a wintering species on the coastal reefs of offshore islets of Papua New Guinea, but I had never seen it in North America. eBird reported the species right where I was looking, but I saw not a single tattler in my searches. Where were the tattlers? Other birders had photographed the species and its nest right there on the shingle flats. The season was right. But the species eduded me.

Further north, near Jensen's Corral (a now-defunct trail-riding operation), was a side road that led to a microwave repeater station at the base of Surfbird Ridge. The well-informed desk officer at the tourism office in Dawson City had recommended that I climb Surfbird Ridge to find nesting Surfbirds up in the high ridgetop scree (scree is loose rocky debris atop a mountain or mountain slope). I made the climb in the early morning, slowly ascending the humpy tundra ridges. Mosquitoes rose out of the vegetation to collect my blood. This kept me moving in spite of the steep incline. The ascent took more than 90 minutes. Once on the ridge crest I began to tour the upland ridge network in search of Surfbirds. This was wild country that seemed to go from crest to crest with no end in sight. Being all alone, I was nervous about losing my way among the network of ridgetops. The sky was blue, and the visibility was perfect over the rocky tundra, but I could no longer see my car or the highway—the only source of unambiguous orientation.

The inaccessible breeding habitat of the Surfbird made finding its nest one of the enduring challenges of North American ornithology. Olaus Murie found a pair with young in 1921 while hiking in the uplands of what is now Denali National Park. In 1926 Joseph Dixon visited this habitat to successfully search for the species' nest and eggs. Only recently, it was found that on the northbound spring migration, the Surfbird stages in Prince William Sound of coastal southwest Alaska—the very place Captain Cook's team collected the first specimen of this species in 1778. This was when Cook was searching (unsuccessfully) for the Northwest Passage—a far northern sailing route from the Pacific to the Atlantic. Because of recent climate change, that route is now open in late summer.

I logged five hours up top, but no Surfbird. Highlights included a Collared Pika squeaking from its rocky scree home territory. This was a new mammal for me, the northern cousin of the American Pika that I had seen on hikes in the Wyoming mountains. I also found an American Golden-Plover on territory. It did a broken-wing display. The bird allowed a close enough approach for me to obtain full-frame images highlighting its stunning plumage. The most common bird on the tundra ridge was the Horned Lark (reminding me of its abundance on the Bentonite Road in Montana). Flocks of Redpolls infested the dwarf birch thickets. My search for another one of the Ridge's rarities, Eversmann's Parnassian butterfly, was unsuccessful. In fact, I saw no butterflies that day. Back down at the base of the mountain, American Robins and Savannah Sparrows were singing from the low shrubbery at the verge of the wet tundra.

Having descended the ridge after my failed hunt for Surfbirds, I drove down to Jensen's Corral where my Dawson City tourism adviser had told me to look for the elusive Smith's Longspur. A Whimbrel called out as it flew by, presumably touring its breeding territory. After a wait, hunkered down in the open tundra, I managed to hear the distinctive high tinkling and chattering series of a male Smith's Longspur, hidden in the low vegetation. This ocher-breasted, sparrow-like songbird is one of the specialties of the park, and the male moved cautiously in and out of its tundra hiding places. My

Surfbird ***Calidris virgata***

Other names: Aphriza virgata

Appearance: The Surfbird looks a bit like a cross between a knot and a turnstone. A recent molecular phylogeny places the Surfbird as sister to the Great Knot (of Eurasia). Its strangely shaped and abbreviated bill is the most distinctive aspect of the bird's morphology. Its breeding plumage is demurely handsome.

Range: This unusual shorebird nests in Alaska and northwesternmost Canada and winters southward along the Pacific coast. Most remarkably, during the boreal winter, Surfbirds are found from Kodiak Island, Alaska, south to the Strait of Magellan, Chile, a distance of nearly 11,000 miles, and the ribbon-like winter range extends inland only a few yards above the tide line. The birds stage in spring in Prince William Sound.

Habitat: The species breeds in bare and rocky tundra atop mountain ridges far from salt water. In contrast, wintering birds frequent wave-pounded rocky coastal shores that they share with Black Turnstones and Rock Sandpipers. Individual Surfbirds can be found on their Pacific coastal wintering range in the Lower 48 for most months of the year.

Diet: On its tundra breeding habitat, the species consumes arthropods and some berries. On the Pacific coast during the nonbreeding season, it takes mainly small mollusks and barnacles, harvested from the wave-washed rocks.

Behavior: The male Surfbird conducts a striking display flight high over his rugged nesting territory, soaring in the wind with trembling wings and vocalizing. Both sexes attend the nest.

Nesting: The female lays four eggs in a sparsely decorated scrape in the ground atop a bare and rocky ridge often frequented by Dall Sheep. The eggs are buff with red-brown spotting.

Conservation: This species is classed as Least Concern by the IUCN, though the population trends downward. A rough population estimate is 100,000.

encounter with the Smith's Longspur rescued a day of ornithological disappointment, having dipped on the two breeding shorebirds. I had now seen all four of the world's longspurs on my shorebird travels.

The next day I departed Tombstone Park at three in the morning and drove northward on the Dempster Highway toward Eagle Plains. The road wound through a series of stark and barren mountain ranges. Mine was the only car on the road, not surprisingly. With the summer solstice approaching, it was not dark when I departed. I dropped down through a pass in the Ogilvie Mountains and entered a pretty lowland spruce woods in the Blackstone drainage. A couple of miles onward, the car flushed a bull Moose, which ambled off into a wet thicket. Then there was a Moose cow and calf posing in a clearing in the spruce forest. Further down the highway, a big boar Grizzly crossed the road at a run. Luckily, the beast crossed from right to left, which allowed me to brake, open the driver's side window, and start shooting with my small lens. I shouted at the bear, which made him stop and turn around to eye me. Then I shot him with my long lens. This big, brown-furred bear had a huge head, small dark eyes, and a big backside. He was massive! A few seconds passed, and the big guy launched onward into the bush, as fast as his feet could take him. His backside rose and fell as he hoofed it out of there.

I arrived at Eagle Plains with half a tank of gas. The stretch of road I had just completed was the longest gas-free stretch on the Dempster. Clearly, my earlier concern about running out of gasoline was unfounded. No matter, the first thing I did was fill the tank. The gasoline was very expensive, but how could it not be, given our isolation? The little hilltop community had an automotive repair shop with a line of waiting customers, a hotel, and a restaurant. I headed in for a big morning meal and to use the hotel's Wi-Fi to catch up with the outside world; it had been days since I had connected with my wife. I tucked into a breakfast featuring three meats and other old fashioned goodies. Also, a big mug of hot coffee. I was still on a high from that Grizzly encounter. I reviewed the images and loved what I saw. Also, the restaurant chalkboard announced that the Peel River Ferry was now open, which meant I could keep moving north.

Back on the Dempster, I stopped every now and then to birdwatch. When I crossed the Eagle River, I recorded American Tree, White-crowned, and Savannah Sparrows; Willow Ptarmigans; and Greater Scaup. In a stand of small Black Spruces, I found an American Robin and a Varied Thrush. Boggy areas up the road were populated with Black Spruce and Tamarack. From the Eagle River the road ascended through foothills toward the spine of the Richardson Mountains. At one high point a large signboard trumpeted my arrival at latitude 66°33′ north—the Arctic Circle. There was something almost mystical about standing on this spot—a universally known global feature that seems so far from home.

Twenty-five miles up road from the signboard announcing the Arctic Circle came the turnoff to the Rock River Campground. This was one of only four formal campgrounds between Dawson City and Tuktoyaktuk. I stopped there for three nights. The campground was nestled in a stand of mature White Spruce and Quaking Aspen fringing the Rock River, which cut a path through the rugged hills. It was a surprise to see these large spruces here, as much of the landscape in this general vicinity was barren tundra. Historically, this river-cut through the hills was an important passageway for the vast Porcupine Caribou herd. Migratory caribou herds are named after their calving grounds, in this case the Porcupine River. Here, in spring and fall, the Gwich'in First Nations people would harvest caribou, an important semiannual source of food and fiber.

I set up my tent and strung up my green rain tarpaulin over the picnic table but made no effort to set up the electric bear fence around my mountain tent. I simply didn't think I needed protection from invasion by large mammals here. That said, I was the *only* person sleeping in a mountain tent, and the tent offered only the flimsiest protection from an intruder of any sort. The other campers were inside sturdy RVs. No matter, I had three peaceful nights of sleep and encountered not a single large mammal at this friendly campground.

From my Rock River base I ranged about on foot in the woods, by bike on the gravel road near the campground, and by car up and down the Dempster Highway. The site offered access to riverine spruce woods, hilly tundra, and rocky scree slopes of the Richardson Mountains. The only mammal I saw in the spruce forest was a Red Squirrel, a species I can find back in Maryland. I spied a small herd of Barren Ground Caribou foraging on high tundra uplands north of the campground. The caribou were moving slowly across the face of the ridge, their ragged pelage blending in with the background browns of the tundra. These were the only caribou I saw along the Dempster Highway. The main body of the Porcupine Caribou herd was summering northwest of here, along the coast.

I started the next morning at the Rock River Campground with a bird walk. The highlight was a handsome yellow-capped male American Three-toed Woodpecker. There were also Spotted Sandpipers, Varied Thrushes, American Robins, Canada Jays, Common Ravens, Swainson's Thrushes, Wilson's and Yellow-rumped Warblers, Dark-eyed Juncos, and Redpolls. The Spotty was even farther north than the breeding range shown in my field guide.

A few miles further, the road rose through Wright Pass, right on the border of the Yukon and Northwest Territories. Parking in the gravel pull-aside, I scanned the tundra hillsides, hoping to glimpse a Wolverine on the move. Then I looked down and saw footprints in the fine gravel of the roadside lookout. To one side of the car were the paw marks of a Grizzly and to the other side were prints from a Gray Wolf. These reminded me of just where I was in the world.

At Wright Pass, I hiked up through some boggy tundra and then up a rocky mountain summit cloaked in blackish scree flecked with variously colored lichens. The summit was free of trees or even tundra vegetation—just a stack of dark lichen-covered rocks. The mountaintop vista was of a dun-colored landscape of rolling tundra hills, with strips of snowpack where the sun had failed to melt the winter's snowfall. It was a barren environment, with no mark of humankind in view but for the ribbon of roadway. The day was cloudy, chilly, and a bit gloomy. A three-hour hike across this barren landscape produced few sightings of wildlife. No pika, no marmot, few birds: Savannah Sparrows, Redpolls, American Pipits, and Lapland Longspurs. No sign of a Surfbird.

From Rock River I headed northward to my destination—the Arctic Ocean. Departing from the Yukon and the unglaciated plateau lands, my route entered the Northwest Territories, a vast area once covered by the Laurentide ice sheet. The wooded lands were dominated by White and Black Spruce, Tamarack, and Quaking Aspen. From the top of the plateau, the vista northeast looked down onto the flat Mackenzie lowlands stretching to the Arctic Ocean. Down the slope was the Peel River, which I crossed by cable ferry. This was Gwich'in territory. I was now near sea level (back in Tombstone Territorial Park, the road rose to a pass of more than 4,000 feet elevation).

At the community of Tsiigehtchic, the Arctic Red River meets the Mackenzie, and the road there crosses the mighty Mackenzie by ferry. Here, of course, I stopped for gasoline, at a convenience store operated by Gwich'in residents. The Mackenzie has cut massive, dark gray cliffs through the spruce-capped sediments. No cable crossing on this broad and imposing river. Flocks of Cliff Swallows foraged out over the river where I crossed.

This put me in the range of the Gray-headed Chickadee, one of North America's rarest songbirds. As I went northward, I stopped at every watercourse crossed by the road and played back the voice of the chickadee with my iPhone to attract this rare Holarctic chickadee, which in this region favors stunted spruce and dwarf willow by stream sides. But I encountered no chickadees of any stripe. I was now northeast of the Mackenzie, which flows northwestward into its vast Arctic delta. A lowland boggy forest of well-spaced Black

Spruces and Tamaracks cloaked lowland flats. The Black Spruces had a yellowish cast; little hillocks were topped by much taller and greener White Spruces, which favor better-drained soils. South of Inuvik, the road crossed a tall, straight ridge that formed the eastern shore of Lake Campbell. I stopped at a pullout and hiked to the top of the ridge—the Tithegeh Chii Vitaii lookout. The ridge was cloaked in White Spruces. The area offered the last of the big trees of my trip northward.

The town of Inuvik (population approximately 3,000) welcomed me at five o'clock in the evening on a holiday weekend. After my splendid isolation, it was a shock to be in such a bustling urban setting. I shopped for groceries and ice for my cooler and then pulled into the Jàk Territorial Park campground for the night. It was noisy, chaotic, and, in fact, at capacity. I was put off by the noise and hassle of Inuvik. Without a sensible alternative, I decided to push on to Tuktoyaktuk. It was another 94 miles on a rather indifferent gravel road with virtually no traffic. This Arctic Ocean coastal community was within the North Slope breeding grounds of the Hudsonian Godwit. Would "Tuk" be another Beluga, with ready access to nesting godwits? Perhaps, but I had no Nathan Senner to advise me.

Once I escaped from the confines of Inuvik, I was quickly out into North Slope tundra flats, filled with ponds and lakes. The sky was gray, and the tundra was dreary and drab but birdy. A pair of Parasitic Jaegers glided over my car. Out on the lakes I saw White-winged Scoters, scaup, and Common Loons (all local nesters). This far north I had hopes of glimpsing a Yellow-billed Loon, but no luck on that front. A thought came to me: it was remarkable that on my 980-mile drive from Anchorage to Inuvik I had not seen a single diurnal raptor from the road. Where were the hawks and eagles? I drove slowly in the gathering evening gloom, being careful because of poor road conditions and hoping to see some interesting birds. Did godwits breed in this tundra? I encountered a Long-tailed Jaeger perched atop a road sign. I saw Northern Pintails and Tundra Swans.

I drove into the coastal Inuvialuit hamlet of Tuktoyaktuk at eight o'clock in the evening, the sun low on the horizon. This Indigenous community marked the end of the road. Beaufort Street took me to a large gravel parking lot set on the verge of the iced-over Beaufort Sea. I was apparently a latecomer, for I found a large array of RVs parked along the fringe of the lot. The parking lot featured a porta-potty (thank heavens) and a large, covered picnic gazebo as well as a fanciful piece of modernist public art celebrating Tuktoyaktuk. This would have to be home for the night.

It had been a long day, with various stops and two ferry crossings, so I immediately started heating a can of soup for supper at one of the picnic tables in the gazebo. At the adjacent table, a husband-wife couple of retirees from Newburgh, New York, was doing the same. We chatted and made friends, and soon we were joined by an array of Inuvialuit residents, interested in getting to know us. Town residents arrived by bicycle, car, and all-terrain vehicle, and soon it turned into a major social event. An Inuvialuit husband and wife made themselves at home at my picnic table and took note of my every move, peppering me with questions. I learned from them that fishing, whaling, and hunting (caribou, geese) were still a major part of seasonal life for the people here. They spoke of drying fish for the winter. I wanted to engage with my local hosts but I was tired and grumpy, and was thinking mainly about getting some sleep.

Another resident arrived with a bottle of vodka in his hand and started rummaging through the back of my open car. Little children wandered about. Cars with noisy exhaust systems came and went. I downed my can of soup and retreated to my car for the night as I did not feel comfortable setting up my tent on the gravel of the parking lot with so much vehicular traffic. Perhaps 20 RVs were parked with me around the rim of the Arctic Ocean parking lot for the night. These folks were cocooned in their vehicles. It was June 17, and latitude 69° north, so it never got dark.

The next morning I rose very early, seeing not a soul, neither from town nor from the nearby parked RVs. A Pomarine Jaeger zipped right overhead. Redpolls and Lapland Longspurs were singing. I also heard an American Robin vocalizing from someone's backyard. The coastal landscape was dominated by sea, bay, lake, and pond, with low peninsulas among the waters. No trees, some low birch and other shrubbery, but mainly water and low tundra. Flat-topped conical hills filled with permafrost, called pingos, were scattered around the interior to the south. I drove around town in search of good birding spots or parkland, but it was wall-to-wall homes and buildings. There was some open water in the protected bays and town lakes, where I saw Glaucous Gulls, scaup, Long-tailed Ducks, and Surf Scoters. Savannah Sparrows were singing. I started back down the highway, moving slowly and making lots of stops to look for shorebirds. I did manage to locate a few—Red-necked Phalaropes, Least Sandpipers, Lesser Yellowlegs, and Whimbrels.

But no sign of Hudsonian Godwits. I feared I had not planned for them adequately. Nathan Senner told me that nesting godwits were few up here, and indeed eBird indicated that they were nesting in boglands further west in the great Mackenzie Delta. I failed to carry out the due diligence on this component of my travels—reaching out to the Canadian Wildlife Service and contacting researchers working on these godwits in the delta. At the very least, I should have chartered a bush plane in Inuvik to cruise low over the Mackenzie Delta and perhaps take a look at the Kendall Island Migratory Bird Sanctuary, west of Tuktoyaktuk. The end of the road, facing the ice-choked Arctic Ocean, was much less than I expected. I was just one more road wanderer, like the others in their RVs, arriving at the end of the road without a proper plan.

That morning, I slowly began my long journey south from the Arctic Ocean, stopping frequently to scan the tundra for shorebirds, but with a feeling of disappointment, in part compounded by the gloomy weather. Add to that a slight sense of alarm when my car's engine light started to shine bright orange on the dashboard. Was I going to be stranded on a lonely gravel road north of the Arctic Circle? I tried to ignore the dash light and focus on the vast stretch of North Slope tundra all around me. I finally saw a couple of raptors—a Northern Harrier and Rough-legged Hawk. The ponds and lakes all hosted waterfowl—Pacific Loons, White-winged Scoters, Green-winged Teal, and Cackling Geese. The dwarf birch in protected spots were leafless, in winter mode. Further south, where I stopped in search of Gray-headed Chickadees, I found, instead, American Tree Sparrows, Northern Waterthrushes, and Blackpoll Warblers. Common Ravens soared acrobatically overhead. Thirty miles north of Inuvik the first spruces of the Far North appeared. Tiny spruces dotted the tundra landscape. I heard another American Robin.

It began to rain on the drive south. Approaching Two Moose Lake, an emergency vehicle, its lights flashing, was parked beside the road; down a steep embankment lay an RV on its side, looking like a beached whale. This was a good motivation to obey the speed limit in a place as isolated as the Dempster. My car's engine warning light had turned itself off (finally, a bit of good news). I took my time, keeping to the speed limit, eventually driving all the way back to my home in Maryland without incident.

On this trip I drove to the most distant point on the North American continent accessible by road and returned home without a single flat tire, car breakdown, or traffic ticket, having racked up 10,929 miles. Like a Hudsonian Godwit, I made my way northward in spring to western Alaska and the North Slope's tundra lands. And like the godwit, come summer, I quickly turned around and began my journey south, making my way southeast to the Atlantic coast, crossing North America twice in a single year. I had spent quality time with the Bristle-thighed Curlew and Bar-tailed Godwit and three species of bears. The achievements had handily topped the disappointments. This was the finest North American solo expedition of my life—one that I recommend to readers with an adventurous bent.

Climate Change on the North Slope

Since 1979, the Arctic climate has been warming, recently at a rate nearly four times the mean global rate—a phenomenon known as Arctic amplification. The causes of the warming are several. One major cause is the reduced reflectivity (albedo) of the Arctic land and sea surfaces. Ice-free sea surface, snow-free land surface, and darkening of the surface of glaciers from the growth of snow-loving microorganisms all contribute to a steadily increasing absorption of solar radiation in the Arctic region. According to a publication by Rantanen and colleagues, various feedback loops amplify warming. Reduced ice cover at the height of the summer melt season leads to increased absorption of solar radiation, which then leads to further annual reduction in sea ice. An "Arctic-mid-latitude linkage" leads to flows of temperate air to the Arctic, producing further warming. Additional (and considerably more obscure and technical) drivers include Planck feedback, lapse-rate feedback, near-surface air temperature inversion, cloud feedback, ocean heat transport, and meridional atmospheric moisture transport. The sum of all the physical influences is rapid climate warming that exceeds measures predicted by existing models. This raises a question: How is this temperature rise changing the ecology of the tundra environments where our shorebirds are nesting? The allied secondary effect of permafrost melting adds to the conservation concern. Can our tundra-breeding shorebirds adapt to these manifold changes to the climate happening over such a short period? Time will tell. ✵

9 Autumn in James Bay

[We] *saw flocks totaling at least a thousand birds pass* [our] *camp on the west side of James Bay, all in less than two hours at midday.... The wing beats were rather shallow but powerful, and the sound of the rush of wings was plainly audible as a flock passed over at their usual altitude of about 200 feet.*

—Hope and Shortt, writing of the Hudsonian Godwit, quoted in Henry Marion Hall, *A Gathering of Shorebirds*

HUDSONIAN GODWITS COMPLETE BREEDING in July, then the adults leave their offspring on the breeding ground and head southeast to staging areas. There, the adults, in effect, have a late summer vacation to put on fat and build up breast muscle in preparation for their long flights to the Southern Hemisphere. The young born that summer remain for several weeks in the place where they hatched. They will then assemble into traveling groups and move to rich staging grounds to prepare for their own southbound autumn migrations.

One important late summer staging site for shorebirds departing their tundra breeding grounds is the Yukon-Kuskokwim Delta in Alaska. The Quill Lakes of Saskatchewan are another; southbound godwits spend at least a month in these staging sites, feeding and preparing their bodies for their overwater migrations.

A third critical staging site for shorebirds, and godwits in particular, is the west coast of James Bay, in Ontario. Thanks to an invitation from Christian Friis of the Canadian Wildlife Service, I was able to join a team and spend 15 days in the late summer of 2019 camped at Longridge Point to assist with shorebird surveys along the shore of James Bay. Shorebird surveys have been conducted there since 1991. In some years, August day counts of Hudsonian Godwit have hit the hundreds or even thousands. I was most fortunate to be able to spend time with this elite team of shorebird-watchers in this special shorebird staging locality, far from the beaten path.

My plan was to drive from Maryland to Cochrane, Ontario, and then catch the Polar Bear Express train to Moosonee, at the very bottom of James Bay. The drive up took a day and a half. Twelve hours driving on the first day got me to a Howard Johnson hotel in Gravenhurst, south of Huntsville, Ontario; on day two I arrived in downtown Cochrane at two o'clock in the afternoon. There I met the survey crew at our assigned motel for the night—the very basic Thriftlodge. The six of us dined at the Ice Hut restaurant and talked birding as we got to know each other.

The next morning, we took the five-hour train ride to Moosonee, courtesy of the Ontario Northland railway. The train makes this trip up and back five days a week, and it is usually full, populated with a mix of Anglo-Canadians and Indigenous Peoples. We loaded our camp gear into a boxcar and then loaded ourselves into an aging but comfortable passenger car. The train had plenty of boxcar space for carrying snowmobiles as well as canoes and other camp gear. It could also carry automobiles tied down on special flatcars. The Indigenous communities of Moosonee and nearby Moose Factory depend on the train for all manner of supplies as there is no permanent automobile road linking these two northern communities with the outside world.

We spent most of our train ride in the dining car, snacking and watching the boreal scenery—stands of Black Spruces in boglands and White Spruces and Quaking Aspens in drier uplands of this expanse of James Bay lowland taiga. Because of the melting permafrost and poor track conditions, the train averaged only 37 miles per hour. The train route to Moosonee followed the valley of the Abitibi River, which flows northward, joining the Moose River. Moosonee, 12 miles upstream (south) of the head of James Bay, has Ontario's only saltwater port, and it services communities on the shores of James Bay and Hudson Bay. Both bodies of water are entirely ice covered for most of each year. Both bays drain northward into the northwestern passages of island-choked Arctic waters that link to the Labrador Sea to the southeast and the Arctic Ocean to the northwest.

After a night at the Waterfowl House in Moosonee, our group of six flew by helicopter north to Longridge Point. The half-hour flight took us up the western shore of James Bay, offering stunning views of boreal wetlands and forests with all combinations of spruce monoculture, bog, open water, and grassy marshland. This was taiga, well south of the tundra country that dominates Churchill, Manitoba, about 750 miles northwest of here.

The helicopter dropped down into some marshy flats not far from a Moose Cree hunting camp—our field team's base of operations for the next fortnight. The camp was leased for the shorebird survey season from a local family based in Moose Factory. Well back from the shore and west of the camp is a "winter road" that links Moosonee to Fort Albany to the north, but this road is not passable in spring, summer, or fall, when the ground is too wet to carry vehicles. Much of the interior country away from the shore is a bogland studded with tiny Black Spruces and Tamaracks. Nor is there boat access to the camp, because the western shoreline of James Bay is too rocky and shallow to provide safe landing. So, for much of the year, the camp is accessed only by helicopter. Our helicopter had traveled from distant Akimiski Island to ferry us from Moosonee to Longridge Point.

We disembarked the aircraft and started freighting our duffel bags and packs to the camp, about a five-minute slog through the marsh. At the same time, the team that had been working the preceding fortnight started ferrying its gear out to the helicopter. Bloodthirsty mosquitoes swarmed out of the marsh grass. This was, after all, late summer in the boreal country of Canada. During our stay, mosquitoes, horse flies, and deer flies hunted us down whenever we were outside. Doing certain chores required a head net and gloves.

The camp comprised one large cook cabin (our operational base) and three smaller sleeping cabins. They were waterproof, bugproof, and functional—perfect for our needs. The camp was set at the edge of a long and narrow strip of White Spruce forest that grew atop an old gravel beach ridge. The "front lawn" offered a nice view out to the marsh and James Bay beyond. There were trails through the woods. The yard was overgrown with shrubs and wildflowers, and it was littered with all manner of stuff that had been left to the elements by the camp's owners—snowmobiles and their parts, tools, empty 44-gallon drums, and stacks of unused lumber.

We all slept in sleeping bags set on mattresses on plywood bunks. Lighting in the cook cabin was fueled by a small tank of propane gas, as was the cooking stove. Laptops and walkie talkies were recharged via a gasoline-powered generator that was turned on from time to time. We were linked to the outside world by satellite phone. There was no running water. Water for drinking and washing had to be retrieved from a small stream in the marsh a five-minute walk from the camp. This was twice filtered to ensure it was potable. An outhouse rounded out the rustic camp infrastructure.

Our team included four paid staff (Doug McCrea, Gray Carlin, Ross Wood, and Angelika Aleksieva) and a single volunteer (me), here for two weeks to provide miscellaneous support to the team.

My payoff was getting to spend time with Hudsonian Godwits on their autumn staging grounds. Our mandate was to carry out strictly regimented daily transect surveys of all shorebirds (in this case both sandpipers and plovers). This was balanced against the various chores that kept the camp running efficiently—water collection, meal preparation, dishwashing, and kitchen cleanup three times a day, as well as burning of trash, data collation in logbooks, and preparation of a daily work report and bird list (the latter to be uploaded to eBird). Even though shorebirds were the focus, the team was bird-mad and generated an exact count of *all* birds seen each day. The team also carried out secondary research activities: mist-netting and satellite-tagging Lesser Yellowlegs; nanotagging Red Knots, juvenile Lesser Yellowlegs, and locally breeding Yellow Rails for tracking via Motus radio receiver towers stationed along their migratory routes; and blood-sampling Nelson's Sparrows. The Lesser Yellowlegs work was part of a hemisphere-wide project examining the movements of this declining species.

No matter the activity list above, the single-minded and intensive focus of the team was on the daily walking of the shorebird survey transects, which were tied to the tidal cycle and which required hours crossing the maze of coastal marshlands, shorelines, and waterways. This regimen ensured that each one of us was kept busy from before dawn to after dusk (which came late in the James Bay lowlands in August). While there was no special focus on Hudsonian Godwits, this species was the most prominent large shorebird that staged here in August and September. So this was where I needed to be in the latter half of 2019. The question was whether I could physically manage it. Each shorebird survey required walking for four or more hours with another of the surveyors. It was a challenge for me to keep pace with my counterpart, whoever that was.

August 2, 2019, at Longridge Point camp was typical: I rose at five o'clock in the morning (in the dark) and did a water-collection run at six o'clock (wearing a head net to fend off mosquitoes).

I then photographed some marsh sparrows (Savannah, Nelson's, and LeConte's). I worked with Doug McCrea to prepare clean water and wash dishes. North of the drinking-water-access trail I heard a clicking Yellow Rail in the marsh grass. After lunch I joined the shorebird survey of Longridge Point—six hours of walking. In addition to 46 Hudsonian Godwits, we found a handful of Whimbrels, a single Buff-breasted Sandpiper, hundreds of the several commonplace sandpiper species, an adult Little Gull in with a large flock of Bonaparte's Gulls, and lots of Sandhill Cranes. The weather was cool and breezy. The shorebird survey walk was exhausting.

Although the walking requirements for the shorebird surveys were punishing (mainly because my counterparts moved so rapidly), this was where I got quality encounters with an array of shorebirds. On August 6 I saw 16 species of scolopacids: 10 Whimbrels, 37 Hudsonian Godwits, 75 Ruddy Turnstones, 20 Red Knots, 5 Sanderlings, 4 Dunlins, 10 Least Sandpipers, 400 White-rumped Sandpipers, 25 Pectoral Sandpipers, 3 Semipalmated Sandpipers, 2 Short-billed Dowitchers, 9 Wilson's Snipes, 2 Solitary Sandpipers, 35 Lesser Yellowlegs, 25 Greater Yellowlegs, and 8 Wilson's Phalaropes. To see these birds, I walked about 10 miles across marsh and creeks, getting back to the camp after nine o'clock at night.

Certainly the Hudsonian Godwit was the flagship shorebird here, as expected. Numbers were low this year, but I did indeed see godwits every day of my stay. Whenever there was a spare moment, I wandered north to Tringa Creek, where the godwits gathered in small parties, foraging in the shallows of the rocky stream bed. My highest godwit day count was 57. I saw a few adult males in full breeding plumage, some males in transitional plumage, and some adult females in transitional plumage, but most in juvenile plumage, with the diagnostic salt-and-pepper spotting upon the mantle. On most days Whimbrels passed overhead, the largest flock being 85. The most abundant shorebird was the White-rumped Sandpiper; these birds stage in flocks of 100 or more. They were quite confiding, and individuals of this species did end up in the mist net, so I got to handle them from time to time. Wilson's Snipes were remarkably common, and we flushed quite a few into the mist nets. Red Knots came and went in decent-sized flocks. Overall I recorded 20 species of shorebirds during my stay.

The White-rumped Sandpiper was commonplace at Longridge Point in August. That was a good thing, because the species was not familiar to me. Having flocks of White-rumps present at the field site allowed me to spend time with them, learning their look and their habits. They resemble the Baird's Sandpiper in size and shape, though not in plumage. In the autumn plumage the adults were very gray and quite plain. Juveniles were more heavily patterned on the back and wings. White-rumps here remained in flocks, which also set them apart from the Baird's, usually seen as singletons. Of course, the most distinctive field mark of the White-rump is the white patch on the back, just above the tail, but this can be discerned only when an individual raises its wings or sets off in flight. The White-rumps are common autumn migrants in the East, whereas the Baird's Sandpipers move through the mid-western sectors of the Lower 48.

Just to give a sense of the season's shorebird abundance, here are the high day-counts for the top 10 species for August–September 2019 from Longridge Point, Ontario: Hudsonian Godwit, 695; Whimbrel, 192; Ruddy Turnstone, 500; Red Knot, 679; White-rumped Sandpiper, 2,956; Semipalmated Sandpiper, 1,732; Least Sandpiper, 376; Wilson's Snipe, 72; Lesser Yellowlegs, 265; and Greater Yellowlegs, 401. Clearly, a season spent at this site provided abundant encounters with these entrancing migratory wading birds.

Autumn Southbound

The southbound migration of North America's shorebirds is very different from northbound migration in the spring. It includes a series of subevents that expand and blur the front end of the process. For many Arctic-breeding species, many yearling birds will only have made a halfhearted attempt to migrate north to the breeding grounds. Many remain on the wintering grounds and make no effort to move northward. Others that do move northward end up stopping "halfway"—dropping out of the race north and stopping to loaf for the summer at good foraging areas in the Lower 48.

These latter birds become the vanguard of the southbound migration in early July. They are joined by failed breeders that did make it to the Arctic but who started south after failing to gain a territory or a mate. The third wave is breeding adults; these birds get to the Lower 48 after staging for a few weeks in Canada or Alaska. Finally comes the fourth wave—the nestlings of the year, who in groups of young somehow make their way southward, stopping to rest and feed along the way. This series of pulses makes the autumn migration far more complex and productive than the spring migration; it lasts from late June to late October. And there simply are more shorebirds on the planet in the fall because of the summer's nesting effort.

Migrating shorebirds always travel in flocks, and they typically set off just before sunset to maximize orientation cues. Flying in formation (as we see fighter jets do) may reduce energy costs, plus group travel probably aids orientation. Flocks rise through the air column, searching for the most favorable winds to assist their sustained flight speed—normally about 30 or 35 miles per hour. Sometimes migrating flocks can be blown off course; a flock of 1,000 Willets was found resting on the sea east of Newfoundland in May 1907, hundreds of miles northeast of their mid-Atlantic breeding habitat.

One of the more remarkable aspects of migration is the rapid reallocation of body mass to different body parts. Digestive organs rapidly grow to allow increased feeding; later, these organs shrink, and breast muscle needed for sustained flight increases substantially. Birds may gain 50–90 percent of their body weight in stored fat and breast muscle just prior to departure on a long southbound overwater flight. The Semipalmated Sandpipers staging at the head of the Bay of Fundy become so fat in autumn that they have difficulty getting airborne for their long flight to South America.

Eskimo Curlews in August were known to add so much subcutaneous fat that it stretched their skin taut. These birds were nicknamed "doughbirds" because when they struck the ground after being shot, their skin split open, baring a mass of pale yellow fat (the "dough"). The Eskimo Curlew became a great table delicacy. As marketable game, it took the place of the Passenger Pigeon, after that hyperabundant species had been wiped out by the same type of unconstrained market hunting (1800–1918). The disappearance of this curlew is remarkable given that there are credible field reports that this species was abundant in North America, with seasonal flocks that darkened the sky. That said, there are professional disagreements about whether the Eskimo Curlew was hyperabundant or whether it simply migrated in very large flocks. A single flock in Nebraska that landed on the ground reportedly carpeted some 50 acres. The tendency of the curlew to migrate in large flocks would have rendered this species particularly susceptible to market hunting. In spring, birds shot in the Mississippi valley were packed into barrels and transported via train to East Coast city markets. The last reliable reports of the Eskimo Curlew date to the latter half of the twentieth century. A spring migrant was photographed on Galveston Island, Texas, in 1962, and an individual was shot by a hunter in Barbados in 1963. The Eskimo Curlew is now believed by most experts to be extinct.

August 8 was productive at the mist-netting station. In order of abundance, captures included Savannah Sparrows; Red-winged and Rusty Blackbirds; Northern Waterthrushes; Least, Semipalmated, and White-rumped Sandpipers; both yellowlegs; and Semipalmated

White-rumped Sandpiper *Calidris fuscicollis*

Other names: Bonaparte's Sandpiper, Bull-peep, White-rumped Peep, White-tailed Stib

Appearance: The White-rumped Sandpiper is a large peep with a diagnostic white rump, mainly visible when in flight. On the ground it most resembles the rarer Baird's Sandpiper. Both feature black legs, very long primary projections, and a short, straight black bill.

Range: The White-rumped is a super-migrator, with many birds wintering at the bottom of South America. The species breeds from northern Alaska east to Baffin Island, Canada. Birds winter from eastern Argentina south to Tierra del Fuego. As with the Baird's, in spring this bird migrates mainly up through the Mississippi valley, often in large flocks. The species is known to make long nonstop overwater flights in the autumn from New England to South America, covering 2,500 miles during its 60 hour flight. In autumn, the White-rumped is more common in the East and essentially absent in the West.

Habitat: The species breeds on low and wet tundra of the Arctic, and during migration and winter it prefers mudflats and shorelines.

Diet: White-rumped Sandpiper's diet on the breeding tundra is mainly small arthropods. In migration and in winter, its diet is mainly tiny marine invertebrates—worms, mollusks, and crustaceans. The species forages by picking and probing mud in shallow water.

Behavior: The monogamous male displays in the air and on the ground when attempting to attract a mate on his breeding territory. The male makes a variety of strange sounds.

Nesting: The nest is a depression in the wet tundra, usually hidden by tundra vegetation. The nest is lined with a variety of plant matter. The four olive eggs are marked with darker colors. The female alone constructs the nest, broods the eggs, and tends the young.

Conservation: The IUCN Red List classes the species as Least Concern, but the species population of 1.7 million has declined by as much as 25 percent since 1980.

Plovers. A juvenile Hudsonian Godwit was nanotagged. In terms of recaptures, the team netted an American Golden-Plover that had been banded at the same site 364 days before. And a Hudsonian Godwit from 2018 was also retrapped. So, these birds are faithful to these staging sites and can orient across long distances and arrive at specific sites at specific times. That presumably benefits the birds, who gain access to locations that offer up reliable and abundant sources of energy-packed invertebrate prey.

I also had several interesting nonshorebird encounters at the camp. The Yellow Rail is a widespread but always-elusive breeder in boggy grasslands in boreal North America. At Longridge, these little rails called on and off in the morning and after dark. Even when a calling bird was within a few feet of me, it remained invisible. A monotonous *tiktik tiktik tiktik tiktik* emanated from the thick marsh grass, and that was the extent of the encounter.

On one shorebird survey we watched two Merlins—small dark-plumaged falcons that hunt birds—chasing shorebirds out along the coast, always an exciting show. Also, a pair of Northern Shrikes tended a nest in a spruce not far from the mist-netting operation; this nest may constitute the most southerly breeding record for the species in North America. On another day a Blue Jay arrived in camp and spent the day; this was one of only a handful of Blue Jay records for James Bay, as this is Canada Jay country, not Blue Jay country! It's exciting to bear witness to species range expansions in real time.

The Black Bear was the featured mammal at Longridge Point. Several individuals, differing in size, wandered the trails and marshlands daily, and occasionally peered in the windows of our cabins or scratched on our front doors. One afternoon, a small yearling bear stood on its hind legs and peered into the window of the women's cabin; it trotted off when it saw me approaching. Near the mist-netting station at the verge of the flats, individual bears foraged at the edge of the spruce woods or out in the willow swale or marsh grass. They ignored us and continued their foraging, often staying in view 15 minutes at a time. They made for good photo subjects. Other mammals our group saw that fortnight included Striped Skunk, Short-tailed Weasel, Red Fox, American Martin, and River Otter. We came upon footprints of Gray Wolves and Moose out in the mud-flats of Tringa Creek. One afternoon, Doug McCrea pointed out the white form of a Beluga whale far out in the bay. Regarding former lifeforms, some of the larger boulders out on the flats were fossil bearing, featuring stromatolites prominent in these Paleozoic rocks.

As the days wore on at Longridge Camp, I wore down, and eventually had to beg off from participating in the long survey walks. They were too much for me (for instance, one had to walk for two hours just to get to the starting point of the West Bay survey). I instead took on chores at the camp (cooking, washing, filtering water) and at the mist-netting station. This allowed me time to focus more closely on the shorebirds themselves. I got to handle several species at the mist-netting operation (Wilson's Snipe, Lesser Yellowlegs, and Least and White-rumped Sandpiper), and from the netting site we also saw lots of free-ranging shorebirds foraging on the flats along the shore. Perhaps most notable was a count of 50 Wilson's Snipes on August 12. On that same day we saw hundreds of Savannah Sparrows, which were clearly on the move. This is probably the most abundant sparrow in North America.

August 13 was departure day for some of us, but the scheduled morning pickup was canceled because of inclement weather. The helicopter came the next day and successfully extracted us; then we camped out at the Waterfowl House in Moosonee because there was no train south on that day. Delays and waiting are always a part of remote fieldwork, just the cost of doing business in the bush. (I had learned to accept this travail from long years doing field work in New Guinea.) On Monday, we rode the train back south, arriving in

Eskimo Curlew ***Numenius borealis***
Other names: Dough-bird, Prairie Pigeon

Once great flocks of this diminutive curlew migrated in spring up through the middle of the Lower 48 and in autumn down through New England and Atlantic Canada on their way south to the Pampas of Argentina. The Eskimo Curlew's nickname "Prairie Pigeon" derived from its abundance, reminiscent of the Passenger Pigeon.

Appearance: This curlew was the size of a Greater Yellowlegs, with a small decurved bill and a dark eyebrow stripe. It was nearly identical in size, shape, and plumage to the Little Curlew of Eurasia.

Range and Habitat: This curlew's only known breeding site was on the tundra of the Northwest Territories, but it may have also bred in northeasternmost Alaska and perhaps in what is now northwestern Nunavut, Canada. In spring it passed northbound from Texas through the Great Plains, mainly west of the Mississippi.

Diet: The species was noted to feed on the Rocky Mountain Grasshopper during that insect's prodigious seasonal outbreaks. The curlews fattened up in staging areas in eastern Canada in the autumn by feeding on crowberries and various marine invertebrates.

Behavior: In autumn the curlew was seen to associate with American Golden-Plovers. The curlew's voice was said to resemble that of the Eastern Bluebird, a sweet trill given in flight: *tr-tr-tr*. When vocalizing, flocking birds in the air could be heard from a distance.

Nesting: The Eskimo Curlew's nest was a depression in tundra vegetation lined with a few decayed leaves. Its four olive eggs were blotched with darker colors.

Conservation: Wagonloads of birds were shot for local consumption and for the East Coast markets. Autumn birds were particularly favored because of the fat stores they gained to power their overwater travels to the Southern Hemisphere. The IUCN Red List treats the species as Critically Endangered (possibly extinct), with a population of fewer than 50. The assumption of most ornithologists is that the last of the species disappeared in the twentieth century.

An adult breeding Red Knot on its upland tundra breeding habitat.

Cochrane at 10:00 p.m., bunking down at the Thriftlodge. The next morning I departed for points east, heading for Atlantic Canada and the East Coast for more shorebird hunting.

The sojourn at Longridge Point camp was memorable for many reasons, not least the privilege of learning from field experts. Doug McCrea, the senior staffer on site, is a world-class field ornithologist with experience from James Bay to New Guinea. When out in the field, I was continually impressed by Doug's ability to identify shorebirds at long distance and his acumen in counting large moving flocks of birds. He was a font of knowledge about Ontario's birdlife. And he was a kind and welcoming host to me, a novice regarding the ins and outs of James Bay natural history.

Ross Wood worked the mist-netting operation and managed the Motus wildlife tracking tower, which records passing birds that carry a Motus nanotag transmitter. Ross was young but highly experienced in the Canadian backcountry. During our days together I would pepper him with all sorts of questions about the natural history of the area, and he always seemed to know about the bird or mammal that I was interested in.

Established in 2014 by Birds Canada, the Motus Wildlife Tracking System is an ever-growing network of towers topped by receivers that can detect nanotags that have been affixed to migrating birds or other animals on the move. Receiver towers have been set up through Canada, the Lower 48, and Central and South America by participating researchers who wish to track the movement of their focal animals. To be recorded, the passing bird must be within signal range, approximately 12 miles. Some 1,800 towers have been deployed in 34 countries (mainly in the Western Hemisphere); 385 species have been tracked, totaling 44,000 individuals (birds, bats, and even tagged flying insects—butterflies, dragonflies, and bees). This is a low-cost collaborative program with more than 700 participating researchers. New towers are added every year as additional researchers take advantage of the growing network. ✵

10 Quebec to South Monomoy

Standing on the eastern shore of Monomoy Point toward sunset … I was enjoying the play of sunlight on the surf.… Suddenly a flock of Short-billed Dowitchers came rippling in from the sea. Nearing the strand, all espied me, mounted sharply and veered overhead at a terrific pace, the slanting sun illuminating their ruddy underparts.

—Henry Marion Hall, *A Gathering of Shorebirds*

After a night at the Thriftlodge in Cochrane, I did some grocery shopping to refill my cooler with food and ice. I stopped to photograph the giant stylized Polar Bear statue not far from my motel, and then I trundled off eastward toward the Quebec border to continue my solo travels in search of godwits and other shorebirds. I had several destinations in my sights: Tadoussac, Quebec, on the north shore of the Saint Lawrence River; Johnson's Mills, New Brunswick, on the Bay of Fundy; the Schoodic Peninsula, Maine; and Cape Cod, Massachusetts, to visit Provincetown and South Monomoy Island. Each offered special birding opportunities in the early autumn.

In traveling eastward from Cochrane, I was tracing the movement of Hudsonian Godwits on their autumnal migrations to

Argentina and Chile. The godwits mainly fly southeast to Atlantic Canada (and sometimes New England) before heading offshore over the Atlantic to fly nonstop to South America. I would be visiting several of their departure sites along the coast to say farewell to these super-migrators for the calendar year.

My first day of travel, August 16, took me eastward to the Ontario-Quebec border and then northeastward to a series of rural Quebecois communities with unfamiliar names (some French and some Indigenous): Poularies, Landrienne, Rochebaucourt, Val-Piché, Lebel-sur-Quévillon, Miquelon, Waswanipi, and Chapais. This was the most northerly automobile route crossing Quebec, and the road was unpaved in some places. The day's drive totaled 450 miles and took eight hours. In eastern Ontario the landscape was undeveloped, with large stands of mature Jack Pine. Things changed upon entering western Quebec. The land was heavily developed for northern agriculture and looked a lot like rural New England—tidy and domesticated. I stopped to get gasoline and snacks at Waswanipi, which is within a Cree First Nations reserve; the very friendly store attendants were anglophone. The road beyond Waswanipi passed through low country dominated by Black Spruce boglands where the trees were darkened by a blackish epiphyte (perhaps a fungus). At this point the last of the radio stations faded away, and I was out of radio contact with the rest of the world.

The route was not terribly birdy. I photographed a small flock of White-winged Crossbills taking water at a roadside ditch. Later, in a wild section of the route, a female Spruce Grouse rocketed across the road from one spruce patch to another. At seven o'clock in the evening I came upon a Red Fox in the road; it was a boldly patterned dark morph—known as a "cross fox"—not the normal all-red variety. I was now at the northernmost point of the route's arc and in deeply boreal country, with Black Spruces dominating in the wetter zones and White Spruces and Jack Pines in the sandy outwash and rocky uplands. I set up camp for the night in Chapais at a campground named Camping Opémiska. As the crow flies, Chapais is about 250 miles due north of Montreal. This was backwoods country and 100 percent francophone, and I struggled to negotiate a campsite with the sympathetic attendant whose English was only slightly better than my French. My hillside campsite overlooked a bay of giant Lac Opémiska.

The next morning I broke camp and followed the only road southeast through central Quebec. I passed the turn off to Chibougamau to the north; with a population of 7,500, it is the largest town in north-central Quebec, developed to exploit gold and copper deposits in the area. This center appeared quite northerly, but the top of Quebec, the head of the Ungava Peninsula, is 900 miles further north. Quebec is an immense province. At around 600,000 square miles, it is about three times the size of France and more than twice as large as Texas. Most of Quebec is tundra and taiga with no roads. That said, the wildest parts of northern Canada are studded with mines of all sorts where diamonds, gold, copper, coal, and other minerals are extracted.

From Chibougamau, Route 167 took me southeast to Réserve faunique Ashuapmushuan, where I set up camp on Lac Chigoubiche. The drive took me through a large swath of boreal logging country. The original forest along the lake shore was intact, but up and down the highway was evidence of past and present industrial logging of the spruce and Jack Pine. Apparently a "reserve faunique" protects wildlife but not timber. But then again, I suspected this area was also heavily trapped for fur-bearing fauna. The name Ashuapmushuan is reported to be an Innu word meaning "where we watch for Moose." Yet sadly, I saw no Moose here.

I spent a day and a half traipsing about in all directions in search of birds—on foot, by bicycle, and by kayak. It was August 18, and the forested landscape was essentially devoid of birdlife; I had never experienced such silence. Yes, I recorded American Crow, Common Raven, Common Loon, and Bald Eagle, but the forest was quiet—no birds in song. Nor did I find non-singing birds. I did a lot of sound playback to draw out songbird flocks, but this attracted only small numbers of this summer's hatchlings—juvenile birds in unkempt drab plumage. I suspected I was experiencing one of the little-studied songbird phenomena of the late summer, when the adult northern breeders head south for a New England holiday, leaving their offspring to grow up and then figure out how to migrate to the tropics and subtropics for the winter. The songbirds may have been imitating the behavior of the shorebirds!

Kayaking out on the lake, I saw only a single Belted Kingfisher. No swallows foraging about. No warblers or vireos singing. The summer migrant landbirds had headed south to richer staging grounds before making their long flights to Central and South America. I had never consciously experienced this before, perhaps because this was most conspicuous in the boreal forest well north of the Canadian-US border, where I had no August experience. This emptying of breeding habitat is a phenomenon that merits further field study (see Zust and colleagues, cited in the bibliography).

From Réserve faunique Ashuapmushuan I followed the course of Rivière Ashuapmushuan southeast to Lac Saint-Jean, in a flat and fertile valley that was open and heavily settled agricultural country, again with a feel of New England (though every radio station was francophone). On the east side of the lake, I headed into the drainage of the Saguenay, which enters a rocky gorge that looks like some place in the Adirondacks. The drive through Saguenay Fjord National Park was stunning, with high forest-topped granite cliffs and a broad and deep blue river that runs 40 miles down to the bucolic tourist town of Tadoussac; there the Saguenay meets the mighty Saint Lawrence—14 miles wide at this point. I set up my tent at Camping Tadoussac, atop a high bluff overlooking the Saint Lawrence, which was fog shrouded when I arrived. The campground was wooded in birches, firs, and spruces.

I visited the office of the Tadoussac Bird Observatory in Les Bergeronnes, about a dozen miles northeast of the campground, to interview the local staff about a songbird migration event that took place in the Tadoussac region so remarkable that it was reported in the *New York Times*. No, this was not about shorebirds or godwits, but it was too phenomenal to ignore.

On May 28, 2018, Ian Davies, a staffer at the Cornell Lab of Ornithology, and five colleagues—one from Cornell and four from the Tadoussac Bird Observatory—counted migrating birds at the Tadoussac dunes, just northeast of town. Over a period of nearly 10 hours, the group tallied 726,383 birds of 108 species. The count did include a few shorebirds: 1 Short-billed Dowitcher, 1 Spotted Sandpiper, 2 Solitary Sandpipers, 1 Lesser Yellowlegs, and 3 Least Sandpipers. The major action of the flight, however, involved the wood warblers. Here are the team's totals for some of the more abundant species: 144,300 Bay-breasted Warblers, 108,200 Cape May Warblers, 108,200 Magnolia Warblers, 72,200 Tennessee Warblers, 72,200 Yellow-rumped Warblers, and 50,500 American Redstarts. These were migrating birds brought north by a southwest wind that struck a system of rain and northwest winds at Tadoussac, which pushed the birds back southwestward along the north shore of the Saint Lawrence. At one point (near noon) some 50 birds were passing the survey point each second!

Even more fantastic, on May 28, 2023, five years to the day of the original event, it happened again. The count was not as high, but more than 250,000 wood warblers were tallied. Clearly Tadoussac is the place to be at the end of May when a southwest wind strikes an approaching cold front from the north.

I was there in August, so I couldn't expect a similar warbler explosion. So I booked a ticket on a whale-watching boat for the next day, and then I went out to the sand dunes to have a look around. I saw only a few warblers, but I scoped large rafts of sea ducks out on the Saint Lawrence.

The next morning broke clear and cool. I headed downtown to get in line for my whale watch. I spent time birding in the harbor and came upon an American Mink venturing out from a little sprucy woodland to raid a collection of trash bins behind a restaurant. While I tried to photograph this urban mink, I turned around and there were Louise Zemaitis and Michael O'Brien, expert naturalists and birding friends from Cape May, New Jersey, standing in the street a few feet from me. They were there on a busman's holiday, scouting out this miracle location for possible future spring warbler visits. They were as surprised to see me as I was to see them. Since I had a morning whale watch to attend, we said our goodbyes and I headed to the boat.

The whale watch was poorly run, the boat was overcrowded, the participants were rude, and the naturalist's narrative was difficult to understand, but conditions were nice, and the whales were abundant—pods of Belugas and also Humpback, Minke, and Fin Whales. Sadly, the boat did not even slow down to enjoy the Belugas (the captain was apparently in a hurry to get to the Humpbacks). The snowy-white Belugas were easy to spot because they traveled in pods and their gleaming arched backs stood out against the dark water when they surfaced to take a breath. The Minke Whales also showed themselves in good numbers, and they approached the boat to allow close-up looks, their smooth, dark gray backs with the handsome hooked dorsal fin small but prominent. Of course, the Humpbacks were the most engaging because of their blowing, fluke display when diving and the occasional breaching, where much of the creature comes out of the water. The single Fin Whale was huge but elusive; one sees little more than a gray arched back and a tiny dorsal fin. Gigantic Blue Whales are present in the Saint Lawrence annually, but none was seen on this day.

That afternoon I headed northeastward up the north shore of the Saint Lawrence to Les Escoumins to catch the ferry across the great river. The harbor was filled with several thousand Black-legged Kittiwakes and Bonaparte's Gulls plus a few Northern Gannets. Spending time photographing the gulls made the wait pass quickly. The crossing of the river took place mainly after dark, so there were no additional whale-watching opportunities. By the time I pulled into Camping Municipal de Trois-Pistoles on the south shore, it was 10 o'clock. A friendly and anglophone attendant got me registered and gave me a very nice camping spot that was well wooded and near to the (free) hot shower and (clean) indoor toilet. Exhausted from the long day, I unfolded my sleeping bag in the car and got right to sleep without eating dinner or setting up my tent.

On August 21 I traveled from Trois-Pistoles, Quebec, to Upper Dorchester, New Brunswick, a drive of about eight hours and 375 miles. I traveled east-northeast along the south bank of the Saint Lawrence to Rimouski, Quebec; at Mont-Joli, at the base of the Gaspé Peninsula, my route turned southeastward and followed the valley of the Matapedia River, a gorgeous and salmon-filled stream in a forested valley reminiscent of the uppermost Hudson River. Lots of fly-fishers wading in hip boots populated the river, doing their thing, along with quite a few flocks of mergansers in female plumage. The Matapedia dumps into the Restigouche River at the New Brunswick border. Eastern White Pines began appearing in numbers as I headed south into New Brunswick. I had not seen this grand conifer north of the Saint Lawrence.

The drive through eastern New Brunswick featured a deeply rural mix of forests logged and unlogged, flower-bedecked pastures, barley fields, tiny towns, and mainly francophone radio stations. The only wildlife were six road-killed porcupines. I lost an hour by crossing borders—it was now Atlantic Daylight Time! Late August is perhaps the high season for this underappreciated eastern province of Canada. It was sunny and warm, and the traffic was light.

My route took me through Amqui, Causapscal, and Milnikek in Quebec and, in New Brunswick, through Campbellton, Charlo, Bathurst, Miramichi, Moncton, and Memramcook, the birthplace of the Acadian people, whose history is both complex and largely tragic. Memramcook was originally a Mi'kmaq First Nations community. The Mi'kmaq people welcomed the Acadians on their arrival from overseas in 1700. Many of the Acadians were deported by the English in 1755, but this town survived. Today New Brunswick is a mix of Indigenous, French/Acadian, and Anglo-Canadian people, and the town names reflect that diversity.

Upper Dorchester sits at the head of the Bay of Fundy, which is famous for its massive tides that can reach 52 vertical feet—the highest on Earth. I was in Upper Dorchester to visit a research colleague, Michelle Venter, who herself was on a family visit from her current home in British Columbia. Michelle and I had done fieldwork in Papua New Guinea measuring carbon stocks in upland forests. Michelle grew up in New Brunswick and was visiting her sister.

The highlight of my brief visit was a field trip we took to Johnson's Mills Shorebird Interpretive Centre, managed by the Nature Conservancy of Canada. The vast mudflats of the Bay of Fundy at Johnson's Mills form a critical staging point for Semipalmated Sandpiper flocks in autumn. Scoping the wet gray-brown flats in the early morning light, we could see thousands of the diminutive sandpipers. Because it was low tide, the birds were scattered and not concentrated in flocks. The Centre staff reported that so far that year the highest day count of the little sandpiper with the partially webbed feet had been 120,000. The adjacent village of Dorchester proudly featured a large statue of a Semipalmated Sandpiper. Recently, the rise of a local baitworm harvest industry on the Bay of Fundy—which involves digging for the worms in the flats—had apparently impacted the sandpiper's ability to feed, substantially reducing the foraging success of the sandpiper on its favored prey—*Corophium volutator*, a marine amphipod. Inadequate prey resources at this critical autumn staging site may be leading to a rise in mortality of these birds on their southbound migrations and subsequently in their far southern wintering sites. This is a species that has apparently exhibited substantial recent population declines.

After saying my goodbyes to Michelle and her family, I headed southwest toward the US border, on my way to Winter Harbor,

Maine, and the Schoodic Peninsula. This was a six-hour drive, covering about 400 miles, all on back roads, passing through River Glade, Petitcodiac, Anagance, Penobsquis, Quispamsis, Saint John (the second largest city of the province), Prince of Wales, and Lepreau. At Saint Stephen I took the little town road to the border at Calais, Maine, to avoid the larger and better-known border crossing with its attendant delays. In Maine, I kept to the back roads down through Meddybemps, East Machias, Machias, Jonesboro, Columbia Falls, Steuben, Gouldsboro, and finally Winter Harbor. Coastal Down East Maine in late August was as sublime as eastern New Brunswick but with slightly less of the timbering/logging flavor. What a pleasure to drive the small roads here in this late summer season, the weedy fields ablaze in deep yellow thanks to the flowering goldenrod.

I arrived late in the afternoon at Schoodic Woods, the brand-new National Park Service campground on Schoodic Peninsula, just east of the more famous Mount Desert Island, and the sign at the front gate read "campground full." But it turned out they had a single small campsite available for a single night. After pitching camp, I drove down to Schoodic Point at the end of the day to find the rocky headlands engulfed in pea-soup fog. It was dark and gray, and visibility was nil. I was thinking about that first Whimbrel I glimpsed at nearby Corea back in 1969. I was not going to see any Whimbrels this evening. Back at the campground it was clear and cool.

After a chilly night in my little tent, I biked down to the now fog-free Schoodic Point early in the morning. Dropping my bike at Blueberry Hill parking lot, I climbed the Anvil, a promontory at the point. In 15 minutes I was on the rocky top with a lovely westward view to the summits of Mount Desert Island. The forest that cloaks the Peninsula was dominated by spruces, and the climb took me through picturesque boreal forest interior, with lovely ground cover of lichens and mosses. It was a surprise to see a few gnarly Jack Pines; they are best known in the sandy outwash of Canadian shield forest far west of Maine. The woods were quiet.

Down on the shoreline, I checked out the waters of the rocky inlets and headlands. This was classic central Maine coastline. Small groups of Common Eiders bobbed offshore. A single Black Guillemot in winter plumage floated nearby. Then, before I got back on my bicycle, I heard a trilling call in the sky, and a Whimbrel dropped down onto the seaweed-covered set of rocks in front of me.

Almost 50 years to the day that I saw my very first Whimbrel at nearby Corea, here was a ghost from my birding past, welcoming me back to the coast of Maine. I happily watched the bird rummage through the wrack for marine edibles. The one jarring surprise this morning was the abundance of colorful buoys marking the location of lobster pots all along the shoreline of the National Park. I was surprised to see these waters being harvested so heavily and so close to the shore of this national protected area.

By midmorning I was on my way to Hope, Maine. Passing by Sorrento, I spotted flats with shorebirds and stopped to scope them—Black-bellied Plovers, both yellowlegs, and a group of Bonaparte's Gulls. No godwits. I was headed to the home of a colleague, John Morrison, in the interior uplands just northeast of the Camden Hills. As I drove up the gravel driveway and over the brow of a hill, I saw a beautiful modern house set in a high clearing, with woodland protecting its flanks and sunlit blueberry fields in various directions. Broad plate-glass windows provided expansive vistas in various directions. John Morrison works for the World Wildlife Fund and is an old birding buddy from his days in DC.

I took lunch on their large sunny deck with John, his wife Serena, and their two boys, Grant and Ward. The house was clearly a labor of love. John assured me there was still much to do to complete this masterpiece, but what I saw was room after gorgeous room with splendid views and myriad artifacts—natural and human made, collected on John's world travels. Now this is the way to live in the woods! We dined and looked out at the beautiful hills to the southeast, and John and I spoke of our explorations in faraway places. John is an expert on the wilds of Alaska and had just returned from a grueling trek across the Seward Peninsula.

After lunch, my GPS led me to Biddeford Pool in southwestern-most Maine, long a site famous for the August stopover of shorebirds (including godwits). I had visited Biddeford Pool several times in the 1970s and wanted to get back for another a look. What I found was a bucolic community on the rocky coast with the "pool" just back from the seashore. I did not find any shorebirds or mudflats (it was high tide), but I wandered around the little lanes and walked out along the shore through a local Audubon sanctuary. Small parties of eiders paddled about off the rocks and gulls soared overhead.

Another hour in the car took me through New Hampshire and into northeastern Massachusetts to tent for two nights at Rusnik Campground in rural Salisbury, just a bit north of Newburyport and Plum Island, home of Parker River National Wildlife Refuge. The refuge is another place famous for shorebirds in autumn. I set my tent under the tall Eastern White Pines and bedded down early in anticipation of a long day birding out on Plum Island.

Early Saturday morning found me at the refuge, hunting for shorebirds. A Hudsonian Godwit had been reported on Thursday, and a Marbled Godwit on Friday. Birders on the Hellcat wildlife trail showed me an American Avocet and several other shorebird species: White-rumps, Baird's, dowitchers, yellowlegs, and a Willet. I yet again saw a Merlin harassing flocks of shorebirds. The afternoon's highlight was a flock of several thousand Tree Swallows swarming over a back bay stand of tall *Phragmites* reeds. Not a shorebird, but a welcome avian effusion. I was still looking for my first Hudsonian Godwit since James Bay. Next stop, Cape Cod.

Early on Sunday morning I drove from Salisbury south through Boston to the South Shore and then onward to Cape Cod, always a great destination for shorebirds, seals, sharks, and whales in late August and early September. I headed to Dunes' Edge Campground just on the outskirts of bustling, festive Provincetown, out at the Cape's very tip. After setting up camp, I drove out to Race Point Beach, on the northern shore of the Cape, to see what was up. I was amazed to see several Humpback Whales blowing simultaneously not far offshore. Seabirds were swarming over them. I scoped Cory's, Great, and Manx Shearwaters in this frenzy. Also, Minke Whales were moving around, breaking the surface and showing their curved little dorsal fin.

The next morning I drove south to Chatham to meet with the US Fish & Wildlife Service's Eileen McGourty to finalize details of my planned visit to South Monomoy Island. As mentioned earlier in the book, Monomoy is one of those historically famous birding locations in New England, much like Plum Island. I would be ferried out there on Tuesday and stay over the Labor Day weekend to be picked up the following Tuesday. When I visited Monomoy Island in 1970, it was a single long and narrow sand island. Today it is three affiliated islands. It's surprising how familiar places evolve!

On Tuesday morning I was loaded into a US Fish & Wildlife Service skiff moored in Stage Harbor in Chatham. This is a gorgeous natural embayment surrounded by sumptuous homes tucked into wooded hills. The surprise of the harbor was finding two adult Peregrine Falcons hunting overhead. They raced by in the bright sunlight of the morning as we slowly wended our way through buoyed boats out into Nantucket Sound.

The skiff zipped me over to the Sound side of South Monomoy, where I jumped out of the boat into the shallows and waded to the beach. This landing place was about a half mile from the historic lighthouse, which would serve as my base for the week. I was then reminded of a peculiarity of the island's vegetation—the sandy trail to the lighthouse was fringed by a profusion of Poison Ivy tangles, the leaves turning deep red with the arrival of early autumn. Some parts of the island were simply inaccessible because of the barriers posed by the overgrown abundance of this noxious native weed.

Otherwise, the vegetation of the island was low, sparse, and benign. A few low pines, plus bayberry, beach plum, cranberry, and blueberry. The wetlands featured Salt Hay, the taller Saltmarsh Cordgrass, and *Phragmites*, among others. Here and there large sand dunes were prominent. No closed forest anywhere. No tall trees anywhere. Several ponds dotted the island; these were frequented by American Black Ducks and Mallards.

The island is a federal wilderness area, and no camping is allowed. My special-use permit allowed me to stay on the island to survey shorebirds. That said, it was perfectly OK for day visitors to boat to the island to hike, surf-fish, and sunbathe. Word is the surf casting from the Atlantic shore is excellent for stripers and bluefish when they are running. There are beaches on both the Sound (bay) side and the Atlantic side. But it is the Atlantic strip of beach that is mind blowing—broad and white, backed by high dunes. This six-mile-long stretch of beach is mainly the domain of shorebirds, gulls, and seals. Few beaches along the Eastern Seaboard have so few human footprints as this one.

By noon on Tuesday I was settled on South Monomoy and had met the Massachusetts Audubon Society bird-banding crew—Nicholas Dorian, Tucker Taylor, and Nancy Ransom. They were scheduled to depart the island the next day, leaving me alone on this "desert island." Before leaving, they planned to brief me on the ins and outs of birding South Monomoy. That afternoon, the banding crew led me to the Powder Hole, where they did their daily count of the waterbirds at this famous shorebird hotspot. It was alive with gulls, terns, and shorebirds. This was where I would be spending a lot of time over the next week. In the evening, we set up scopes with the setting sun at our backs, and we checked out every bird. This count would go into the eBird system via Mass Audubon. The banding team did this evening survey daily, after they had finished mist-netting and banding birds around the lighthouse. My own week of surveys focused on the Powder Hole and the two beach fronts—sound and ocean. Much of each day was spent walking.

Monomoy's three islands are North Monomoy, Minimoy, and South Monomoy. The southern "bulb" of South Monomoy holds the old lighthouse and the lighthouse-keeper's house. This was where I slept and ate. This recently restored facility no longer provides a warning light for passing ships, but instead is used as a research base for the likes of the Mass Audubon bird banders and me. The lighthouse was built in 1823 when the sandy barrier strip was contiguous with mainland Cape Cod and then known as Sandy Point. Sandy Point became an island in 1958 when a spring storm carved a channel between the Cape's "elbow" and what is now North Monomoy.

I had come to South Monomoy because, historically, Monomoy has been the "last stop" for southbound migrating Hudsonian Godwits in autumn. Not as many godwits seem to stop over on Monomoy as in past decades, but I wanted to experience once more this important place in the godwit's story. So I was out looking for godwits every day. First, I came upon a single Marbled Godwit that had settled into the Powder Hole. A couple of days later, I watched as a juvenile Hudsonian Godwit dropped down into the Powder Hole, a half-century after I saw my first Hudsonian here on Monomoy. This paralleled my experience with the Whimbrel on Schoodic Peninsula the week before. It exemplified the splendid timelessness of birding in very special places.

This godwit, in its speckled-backed juvenile plumage, may have gotten separated from its migrating flock of yearling birds heading south from the breeding ground. It spent three days on the island before disappearing, perhaps on a lonely solo flight south across the Atlantic to South America. Or perhaps it joined a flock of vocalizing juveniles that passed overhead one night, or a mixed flock of southbound shorebirds. The southbound migration of juvenile birds remains one of the mysteries of shorebird migration.

During my stay on South Monomoy, there were plenty of other shorebirds to study, count, and photograph. The Monomoys remain a great stopover site for these southbound migrants. The Powder Hole was a favorite feeding and loafing site. As I stood amid the shorebirds there, the cacophony was a "surround sound" of Semipalmated Plovers, Semipalmated Sandpipers, and six or seven other noisy species. I managed to see all the shorebird species I had first observed here in 1970, which was a source of nostalgic satisfaction.

Sanderlings flocked in to forage in the Powder Hole, but it is South Monomoy's Atlantic beachfront that is this species' favored feeding ground. All that white sand! All that wave action! Flocks of postbreeding Sanderlings scattered themselves up and down the six miles of beachfront, hunting mole crabs in the wet sand from dawn to dusk. The back and forth of the birds tracking the in and out of the breaking waves is mesmerizing for beachgoers as well as birders. The Sanderling is the king of this solitary beach for much of the year, foraging at the waterline for hours on end, and loafing and roosting up above the high-tide line among the riprap.

Monomoy is reputed to host a small population of White-tailed Deer, but the only evidence of the species I saw was a single discarded rack of antlers. Two creatures did appear commonplace—Garter Snakes and Fowler's Toads. One morning I walked north on the Sound side of the island in search of shorebirds. It was very peaceful; the shore was littered with the discarded shells of seasonally shedding Atlantic Horseshoe Crabs, but I saw few birds of note. Walking back southward on the Atlantic side was a bit more exciting—windy and sunny on this broad white-sand beach. A single Humpback Whale blew offshore, and an Ocean Sunfish waved its pectoral fin at me. The Ocean Sunfish, the world's largest bony fish, tends to float pancake-like flat on the surface of the sea. The fin it raised first made me think "shark," but it was waved about willy-nilly; a shark's dorsal fin can't do that.

Sanderling ***Calidris alba***
Other names: Bull Peep, Whitey, White Snipe, Beach Snipe

Appearance: This adorable species—for most beach lovers the prototypical shoreline sandpiper—has bright rust-colored breeding plumage, but this is rarely seen except when birds are approaching their breeding territories in the Arctic. Most of the time the birds are seen in flocks on coastal beaches wearing their pale nonbreeding plumage—the adults gray-backed, and the juveniles with backs heavily marked with black speckles.

Range: The Sanderling breeds in Arctic Canada, Greenland, and northern Siberia. North American-breeding Sanderlings winter southward on the three coasts of the Lower 48. The species also winters widely into the Southern Hemisphere.

Habitat: The Sanderling breeds on dry ridges in Arctic tundra. Migrants and wintering Sanderlings favor beaches and coastal tidal flats. In July and August, adult Sanderlings in North America start moving south toward their wintering grounds. Sanderlings are major wanderers, with remarkable abilities to travel. The species migrates and winters in flocks. Most birds in their first year of life do not travel to the Arctic to breed.

Diet: The Sanderling's diet is mainly mole crabs, other crustaceans, and additional invertebrates.

Behavior: The adult male Sanderling carries out a low display flight with aerial fluttering and attendant vocalizations for the female. Most of the year the species is very sociable and can be found in foraging flocks that move up and down sandy shorelines.

Nesting: The nest is a shallow scrape placed on a low gravelly hummock. The female lays four olive-green or pale brown eggs, sparsely spotted with brown and black. Serially monogamous, the female may lay one to three broods a season.

Conservation: The species is categorized on the IUCN Red List as Least Concern. The global population is approximately 700,000 but has declined by as much as 40 percent since 1980.

Sea ducks rested in flocks off the shoreline. Common Eiders were around, as were all three species of scoters, Blacks being the most abundant. Double-crested Cormorants were commonplace.

Gray Seals are plentiful around Monomoy. The seals patrolled the shoreline, often popping their heads out of the water to look at me. Groups of seals just offshore followed me, watching, as I walked the beach. They are curious souls. One early morning I came upon a large group of seals hauled out on the beach. The group included big brutish males, smaller gracile females, and little ones of the year. Among these groups were some with major gashes in their flanks. The result of shark attacks?

I spent more time at the Powder Hole flats than any other place. There were always shorebirds, gulls, and terns at the Powder Hole, so it was always productive. I hunted for Baird's Sandpiper among the flocks of Semipalmateds—like hunting for a needle in a haystack.

There was not much evidence of songbird movement. A few Bobolinks called while passing high overhead, but the warblers seemed few. Perhaps it was still early in the season and these birds were still loafing and fattening up northwest of here in the interior of New England. Tree Swallow numbers continued to build up, their flocks alighting on the dunes. It was fun to watch the swarms that swirled about in the freshening winds. Thousands! One morning an adult Parasitic Jaeger appeared overhead. It circled me and then several times dive-bombed a passing Northern Harrier. Encountering a sea-loving jaeger over land in the Lower 48 is always a treat.

South Monomoy has continued to evolve, as winds, currents, and storms move sand and water in unexpected ways. The good news is that the lighthouse and its outbuildings are under no threat of being washed away. In fact, it appeared that the south end of the island is continuing to grow. The bottom of the island had expanded into a large, rounded lobe, gathering sand carried south down the eastern face of the Cape.

The Lesser Peeps

1. Least Sandpiper *Calidris minutilla*

Other names: Mud-peep, Meadow Ox-eye, Green-legged Peep

2. Semipalmated Sandpiper *Calidris pusilla*

Other names: Sand Peep, Bumble-Bee Peep, Hawk's Eye, Oxeye, Black-legged Peep

3. Western Sandpiper *Calidris mauri*

Other names: Western Semipalmated Sandpiper, Peep, Bumble-Bee Peep

These three North American peeps (diminutive sandpipers) are commonly encountered and can be tough to identify.

Appearance: The Least is the smallest and is droop-billed, brown dorsally, and yellowish legged. The Semipalmated has a short, undrooped black bill, black legs, and a rounded head. The Western has a longish, tapered, and drooping bill, flattened crown, black legs, and in some plumages a distinctive chestnut shoulder patch.

Range: Our three peeps breed in the North and winter from the southern United States to South America. Any of these birds can be encountered in large numbers in migration. They breed in tundra and other tundra-like bog habitats, and they migrate and winter on coastal flats, the species sometimes mixing.

Habitat: They are typically found in flocks, small or large, foraging on mudflats or on the verges of wetlands.

Diet: The diet of these peeps on the breeding ground is mainly insects and small crustaceans. In migration and winter, the diet is a wide mix of invertebrates. The peeps feed by pecking and probing.

Behavior: The monogamous males conduct a vocal aerial display.

Nesting: The nest is a depression lined with plant material, often under or next to taller woody growth. The female lays four eggs that are either whitish or buff and marked with darker blotching.

Conservation: The Semipalmated (population 2.3 million) is classed by the IUCN as Near Threatened and is included on the NABCI Watch List. The Western (3.5 million) and Least (700,000) are classed as Least Concern by the IUCN.

On the early morning of September 3, anticipating a mid-morning pickup and my return to the mainland, I headed down to the Powder Hole for one final encounter with my beloved shorebirds. On the walk down, several Bobolinks passed overhead, giving their *bink* notes. Things were quiet at the sandflats. No godwits; no Whimbrels. Just yellowlegs, dowitchers, Stilt Sandpipers, Willets, and peeps. Nothing new had come in overnight, but a few species had departed. There were plenty of peeps still around, swarming over the flats.

The peeps encompass a selection of the smaller *Calidris* sandpipers. The term "peep" is a bit vague; defined broadly, it includes, from smallest to largest, Least, Western, Semipalmated, White-rumped, and Baird's Sandpipers. The field guides typically omit the last two largest of the group. But for birders, these five are pretty much alike and work well as a group. When they are all together, it is easy to see the size differences, but when you see solitary individuals, it is difficult to determine relative size, even of the largest two (White-rumped and Baird's). Distinguishing the species of peeps in foraging flocks is one of those birding tasks that some love and others hate. It is always best done with a nice spotting scope, which allows the observer to detect leg color, bill curvature, and primary extension. Learning to identify the peeps is much like learning to distinguish the fall-plumage wood warblers. This takes the birder to the next level of capability and is a source of some birding pride.

Back at the lighthouse, it was time to pack up, clean and straighten the house, and start freighting my stuff out to the landing on Nantucket Sound. I used a wagon with balloon tires to move the heavier and bulkier stuff across the Poison Ivy–laced landscape. Hurricane Dorian was approaching. The US Fish & Wildlife Service people wanted me off island before the storm surge hit. By one in the afternoon I was back in Stage Harbor—as always filled with birds—Common Terns, Double-crested Cormorants, several gull species, and an Osprey.

With time to spare and the storm still well offshore, I made a last-minute plan to charter a Cessna out of Chatham airport to search for Great White Sharks off South Beach. At four o'clock I boarded the little Cessna 172 with Tim Howard at the rudder. The two of us took off without delay and within three minutes we were tracking northward on the Atlantic side of South Beach, Chatham to the southwest of us. And there down below in the pale green water was the dark form of our first Great White. Tim banked hard to allow me to photograph the beast, which was less than 100 yards offshore, in shallow water. The white-sand sea bottom made it particularly easy to see the shark's dark dorsal surface as it coursed gracefully through the water. Who would have thought it would be so easy to glimpse this iconic megapredator? These big sharks were here because of the abundance of Gray Seals up and down the shore. And the seal's abundance was because the Monomoys offer a safe and productive wintering haven for them. Tens of thousands winter on the Atlantic beaches of North and South Monomoy. The return of the seals and the arrival of the sharks is a relatively recent phenomenon. Neither were here in the 1970s. Decades of conservation allowed the seals to rebound after their decimation during the whaling years, in which all marine mammals were ruthlessly harvested for their meat and oil.

During our 40-minute flight, we spotted seven Ocean Sunfish floating flat on the surface of the sea, waggling a pectoral fin in their peculiar fashion. And we found three different individual Great Whites on our tour, which took us north almost to Truro, near the northern tip of the Cape. Tim told me that finding Great Whites from the air is easy during this time of the year, so long as the wind is down and the water relatively clear. The best time of all is the early morning when the sea surface is glassy and the sun is in the east.

After returning to the airfield, it was back to Provincetown for the final night of this six-week field trip. I dined at the Canteen in bustling downtown P'town with whale naturalist Dennis Minsky, who had organized a whale watch for me the next morning.

Early Wednesday morning I packed the car and headed out to Race Point Beach to take a last beach walk and to see what was happening out over the water. Not much. I glimpsed one or two distant blows from Humpbacks, but no big collections of seabirds. A single jaeger swooped about in search of a tern or gull to harass.

By midmorning the *Dolphin 11* had passed out of Provincetown harbor with a decent crowd of anxious whale watchers. A friendly threesome of birders from Michigan adopted me, and we spent the trip together, watching birds, fish, and whales.

That morning the seabirds were few: 15 Great Shearwaters, 1 Sooty Shearwater, 1 Leach's Storm-Petrel, 5 Red-necked Phalaropes, and 5 Northern Gannets. Most of the Great Shearwaters we saw right in the mouth of the harbor. The open ocean was relatively bird-free. But the phalaropes are, indeed, shorebirds, so yet another summer day was spent with my target group, albeit one of the oceanic oddballs of the family. It was strange to see these small whitish shorebirds in tight flocks bobbing on the surface of the sea, then rising and flying off as the boat approached. The boat passed several Ocean Sunfish. Our whale encounters were all the Humpback variety. We spent quality time with two mother-and-calf pairs. The calves were pretty much grown up, but still associating with their mothers. One mother did a partial breach with a tail smash, raising a white explosion of water and gasps from the crowd.

This splash ended my six weeks on the road in search of godwits, other birds, and nature. Back in Provincetown harbor, my GPS pointed my car toward Maryland, which after nine hours would end a long summer trip of 3,950 miles.

When I started this series of journeys in 2019, I really did not know where my travels would lead me, though I was confident I would learn something substantial about my target species, the Hudsonian Godwit. I was not seeking to "do" science, just simple natural history, learning about a little-known migratory bird that crosses our continent twice a year in great leaping flights. To do that, I needed to catch up with these birds, in their flocks, when they touched down to take a rest. It was a bit of cat and mouse until I made my way north to the nesting grounds of these lovely birds.

As I traveled, I first learned that Hudsonian Godwits depend upon the abundant agricultural fields scattered across America's midsection in their spring migration. They favor wet spots in bare-earth fields for resting and feeding, not some park or protected area. Thus, they are very much a part of our rural landscape, though they tend to be overlooked by local residents, lost in the vast spaces of our heartland interior. I also learned that, on their breeding ground, these godwits are lovers of boreal bogs, with nesting habitats scattered across the northern tier of the continent. These boglands are filled with sound and life in late spring as the last of the snows melt away. Spending time with godwits where they nest was a highpoint of my journey, and being able to share this experience with biologist Nathan Senner and his team made the experience all the more edifying—on the front lines of the research world. Being able to share the excitement of discovery with field scientists doing their thing is a special treat for the naturalist.

What about being alone out in the countryside week after week? What was that like? The reader might imagine I grew lonely or bored. Sorry, not so! There was too much to see, hear, and react to, day by day and night by night. I dearly love my wife, Carol, but she was at home, engaged with her work, and I was doing what I most enjoy—wandering North America's back roads in search of nature. There are so many enchanting green spaces scattered across the United States and Canada! A naturalist wants to visit them all.

Was I ever frightened? Perhaps not frightened but once I was hyperalert. That was when a large supercell caught up with me when I was driving on the Kansas prairie. The sky blacked out, the wind rose, and the rain pelted down at an angle. I wondered whether there was a twister buried inside that black veil of mists. I pulled under an overpass and turned off the car. My Xterra rocked violently and the wind screamed. Seconds passed like minutes. Then the world quieted down, and I emerged intact. Just a fierce prairie storm. I entered a nearby town to get lunch, and the main street was strewn with leaves and branches, remnants of that passing supercell.

By the time I pulled into my driveway after the final leg of this life-changing experience, I felt I had gotten to know my target bird, but, also, I had gotten to know so many other little-known species and hidden places far from home—natural and beautiful places populated by colorful and remarkable birds and mammals. These experiences are the major payoff to the naturalist willing to get out on the road and follow shorebirds in migration. Experiences such as glimpsing a male Buff-breasted Sandpiper conducting his open-wing display to watching females in a field in eastern Nebraska. Seeing a big old American Badger hump across a back road in Kansas. Being stunned by a glorious flock of 249 American Golden-Plovers hunkered down in a bare-earth roadside pasture in South Dakota. Watching a male Hudsonian Godwit do his noisy display flight over the fen in Churchill, Manitoba. Gaping at a big mother Polar Bear out on the sea ice in downtown Utqiagvik.

There's nothing quite like being alone, out on the open road in the height of spring, traveling a back road to a little-known corner of North America where nature is doing something special. Tracking Hudsonian Godwits allowed me to do this, day after day. Each day offered up new surprises and taught me lessons about the shorebirds. Reading about a bird is one thing. Watching it do its mating display on its breeding ground is another thing entirely.

Dear Reader, I encourage you to get yourself out on the road and see what you can find! ❋

PART III
UNTIL NEXT YEAR

A pair of Buff-breasted Sandpipers forages on a plowed field in eastern Nebraska.

11 Wintering Shorebirds

The afternoon sun leans its rays into the repose of the marshes, when suddenly one of those tremendous floods of life surges over them, sweeping down in the distance like a cloud detached from the sky, an invasion of Valkyrie with all the wild discipline and exultation of speed and none of the menace and terror.

—H. J. Massingham describing a Dunlin flock in Forbush and May, *A Natural History of American Birds*

MY JOURNEY FOLLOWING THE SHOREBIRDS on their migration north took me through the spring and summer months. I got to see these birds on their breeding grounds, and I witnessed how they behave during this important time of year. But that's only part of their story. What are these birds doing during the northern winter? Recall that the breeding season for northern shorebirds is brief, and much of the year is spent away from the breeding ground; for many species, more time is spent in the tropics or even in the Southern Hemisphere.

Although we often speak of "wintering shorebirds," most of our migrant North American shorebirds travel long distances to live their lives in perpetual spring and summer. And this means these birds spend much of each calendar year in the Southern Hemisphere (roughly from September to April) or in the tropics, away from chilly boreal winters in temperate North America. These birds chase benign climates and avoid the harsh winters that less mobile species must endure. Enjoying two warm seasons a year is the great payoff for two long migrations a year. Shorebirds that each autumn head south over the Atlantic from New England are traveling to a Southern Hemisphere vacation spot where they spend as many as seven months of the year loafing, fattening up, and preparing themselves for a race northward when the promise of the next northern spring is signaled by changing sunlight and the bird's internal alarm clock.

"Wintering" is a messier and more variable process, geographically, than breeding, because loafing is much less demanding than producing a batch of offspring. Examine the eBird occurrence map for the Hudsonian Godwit for December to February for the past 10 years, and you will find records of the species scattered from Connecticut and Kansas south to Tierra del Fuego. Looking at the eBird map for the same period for the Whimbrel, records range from the southern tip of South America north to British Columbia and Massachusetts, including many reports along both coasts of the Lower 48. It seems that wintering birds spread out and find a wide range of suitable wintering sites based on age, experience, and inclination. Of course, over that seven-month nonbreeding period, these birds are not tied to any one place; they can and do move about, driven by weather and availability of productive feeding grounds.

Wintering on the Northern Coasts

Some North American shorebirds winter in a true northern winter clime. Here we examine four shorebirds that winter along the East Coast from Maine to the mid-Atlantic. One is our familiar Sanderling, populations of which winter from British Columbia and Maine south to Tierra del Fuego—an expanse of more than 100 degrees of latitude. This is probably the second most widely wintering shorebird on Earth as measured by distance from north to south (after the wintering champion, the Surfbird). Sanderlings spend the northern winter in flocks along the sandy beaches of the Atlantic, Gulf, and Pacific coasts. Flock size generally increases as winter proceeds, perhaps in part because roosting in large groups can provide thermal benefits through cold nights. Since wave-driven coastal beaches rarely freeze up, these birds can continue to forage for sand-dwelling marine invertebrates until spring arrives. Thus, their behavior changes little from autumn through winter to spring. Why some Sanderlings winter in the Southern Hemisphere and others winter in the frigid north is unknown. Perhaps some breeding subpopulations have evolved to winter in the far south simply by chance. The Red Knot is the only other beach-loving sandpiper that exhibits a similar wide latitudinal range of wintering.

The Dunlin, a lover of northern winters, is common in winter from New England and British Columbia south to the eastern and western Mexican coastlines. Dunlins are invariably found wintering in flocks, predominantly on intertidal flats. Their diet in this season is primarily marine invertebrates. In late autumn and early winter, large flocks of Dunlins are found in foraging and loafing sites at coastal estuaries that provide a range of habitats for both feeding and roosting. These flocks are regularly targeted by passing Merlins and Peregrine Falcons, and the sight of a tight, wheeling flock of Dunlins being chased by a falcon is a familiar autumn spectacle. As with wintering European Starlings, the Dunlin flocks move as tight, swirling murmurations to stymie the attacking falcons. The flock in flight, struck by low sunlight, is one of ever-changing shape and color, flashing quickly from dull gray to bright white as the birds show first their backs then their bellies. In the mid-Atlantic shore in winter, the Dunlin is the most abundant shorebird. And for the Dunlin, raptor predation is the single greatest source of population loss in winter.

Our third winter-loving shorebird example is the Marbled Godwit, one of the Magnificent Seven. This short-distance migrant winters mainly on the Pacific, Atlantic, and Gulf coasts of the Lower 48, with

Dunlin *Calidris alpina*

Other names: American Dunlin, Red-back, Red-backed Sandpiper, Lead-back, Fall Snipe, Brant-bird, Crooked-bill, Simpleton, Winter Ox-eye, Little Black-breast

Appearance: Perhaps our most abundant shorebird, the Dunlin is larger than a peep and has a drooping bill. On the species' tundra breeding grounds, this is a handsome sandpiper, marked with a black belly and rusty mantle. Winter birds are very plain.

Range: The Dunlin summers in western and northern Alaska, Arctic Canada, and Arctic Eurasia. This species winters southward from the north temperate zone coasts into the subtropics. Some Eurasian birds winter as far north as Greenland and Iceland.

Habitat: The Dunlin breeds atop low ridges in wet tundra. It can be found in large flocks roosting on sandy flats in autumn, when most shorebirds have retired to warmer climes. Migrants and wintering birds are most commonly found along the coasts, favoring muddy pools, mudflats, and tidal waters, where birds feed with their bellies in the water.

Diet: The Dunlin's diet is clams, worms, insect larvae, and amphipods. It often forages in deeper water, using its long bill to probe the bottom for invertebrate prey.

Behavior: The species is monogamous and sociable; it is often seen in large flocks in winter and on migration. Its voice is a harsh, reedy, and rasping *pjeeev*.

Nesting: The nest, hidden under tundra vegetation on a low hummock, is a shallow scrape lined with bits of vegetation. The female lays two or three olive eggs with brown blotches. Both adults incubate. The adults linger on the breeding ground with their offspring.

Conservation: The Dunlin is on the NABCI Watch List and has been in decline since 1980. That said, the species' current population in North America is estimated to be 2.15 million birds.

some populations edging as far south as Central America. In the mid-Atlantic, small numbers of Marbled Godwits winter in New Jersey and the Delmarva Peninsula, concentrating in productive estuaries with extensive mudflats. Here the godwits roost in small parties along with other flocking shorebirds. Presumably these mixed roosting flocks are an antipredator strategy—more eyes on the skies. When an individual bird needs some shuteye, it can be confident that other individuals in the large flock will be awake and wary, ready to sound an alarm at the sight of a Peregrine overhead.

The Purple Sandpiper is the East Coast's most rugged winter lover. It breeds in the Arctic and shifts southward only hesitantly, appearing on the mid-Atlantic shore in late October and early November. Small parties station themselves on seawalls and rock jetties that are encrusted with seaweed, which the birds pick over in search of tasty morsels. Never abundant, Purple Sandpipers are nonetheless reliably present in their favored foraging locations, which tend to be few. Certainly, this species can be considered our hardiest shorebird. In the thick of winter, Purple Sandpipers subsist on various invertebrates and algae. They stoically work the wet surfaces of jetties, regularly getting showered by chilly incoming waves. Most remarkably, these birds will feed at night in the dead of winter. Presumably this habit is driven by tidal cycles along with the need for additional calories in the height of winter.

Winter Flocking

In winter, one often sees several species of shorebirds foraging together in a single area with minimal conflict. Thanks to differences in body size, sensory modality, leg length, bill size and shape, and foraging method (visual vs. tactile), different species can use different subsets of the habitat to avoid direct competition and conflict when they feed as a group.

Whereas shorebirds during the breeding season tend to be asocial or solitary when foraging, during other seasons some species become hypersocial, foraging in tight flocks. This is certainly true for the dowitchers and the Red Knot. Because these birds are hunting for buried prey, closely adjacent congeners (members of the same genus) will not disturb each other's foraging. By contrast, a visually hunting and picking species, such as the Buff-breasted Sandpiper, will feed in dispersed flocks, because nearby congeners might flush potential prey. Still, foraging in flocks may offer benefits. One bird's knowledge of prey concentrations may guide a flock to a prime site.

As mentioned above, flocks also provide visual defense against aerial predators, especially falcons and bird-eating hawks. And flying in dense flocks seems to offer group protection against an aerial predator strike on an individual. The synchronized and rapid movement of a large flock of shorebirds is a remarkable thing to witness, especially when the swooping flock is attended by a hungry Peregrine. When hunting Merlins are present, Sanderlings abandon their linear beachfront feeding territories and stay in tight flocks. Studies show that predators have more success taking prey from smaller flocks rather than from larger flocks. Despite such flocking strategies, raptors (especially falcons) take substantial numbers of shorebirds as prey each year.

When traveling in flocks, shorebirds still do compete for favored or more productive feeding sites. Among flocking Sanderlings in Delaware in the postbreeding season, aggressive postures and chases among adjacent foragers are commonplace. Even though these birds forage in flocks to foil predators, individual birds are not particularly friendly to each other. Ruddy Turnstones establish dominance relationships between individuals in sites where they regularly forage, and fighting among foraging turnstones is commonplace. At moderate prey densities, individual Sanderlings defend a specific foraging territory, whereas when densities are very low or very high, these birds feed in groups with no evidence of territorial exclusion.

Purple Sandpiper ***Calidris maritima***

Other names: Rock-bird, Winter Rock-bird, Rockweed Bird, Rock Plover, Rock Snipe, Winter Snipe, Winter Peep

The Purple Sandpiper is a strict habitat specialist that demands birders visit a rock jetty in the cold of winter to see the bird.

Appearance: In winter the species sports its drab and dusky-streaked plumage; its legs and the base of its bill are dark orange. It is a chubby and medium-sized *Calidris* sandpiper. Its heavily patterned breeding plumage is rarely seen in the Lower 48.

Range: It breeds in Arctic Canada, Greenland, and Eurasia. American birds winter to the East Coast of the Lower 48 and the rocky shores of Atlantic Canada.

Habitat: The Purple Sandpiper nests on northern and barren low, mossy tundra in stony clearings and on low ridges. It winters in small parties on rocky shorelines and breakwaters along the Atlantic coast in association with Ruddy Turnstones and Sanderlings.

Diet: This sandpiper's summer diet is a mix of arthropods and plant matter. Its winter diet is mainly mollusks and crustaceans harvested from wave-washed rocks.

Behavior: The male carries out a fluttering aerial display high above the nesting territory. The Purple Sandpiper's flight call is a rough and low-pitched *kweet*. It is a monogamous breeder. The species is a capable swimmer.

Nesting: The nest, set in open tundra, is a shallow depression of lichens and mosses. The female lays three or four olive or buff eggs heavily marked with rufous brown. Males do most of the incubation and solely care for the hatchlings.The species moves south late in autumn, staying on the breeding ground with their young longer than most shorebirds. The young of the year migrate south from the breeding ground with the adults.

Conservation: The IUCN Red List classes this species as Least Concern. The estimated population in 2015 was 250,000.

Wintering in Florida

The coast of southwestern Florida is riddled with barrier islands and beaches that are blessed with clement weather year-round, making these habitats ideal for loafing shorebirds at any season. First take note that Florida welcomes all sorts of boreal-breeding shorebirds in summer—species such as Sanderling, Ruddy Turnstone, Marbled Godwit, and Whimbrel. These summertime loafers are nonbreeding slackers (mainly young birds) that skipped the trip to the breeding grounds during the nesting season. These slackers presumably do not make the effort to reproduce because the competition on the nesting ground is fierce, and the winning strategy is to delay and grow physically before joining the competition. But winter is the high season in Florida for these shorebirds, which appear to be slacking once again by staying in Florida rather than doing the full southbound migration to the Southern Hemisphere.

The Gulf coast beaches are good birding in winter. The season's avifauna includes shorebirds, gulls, terns, cormorants, frigatebirds, herons, and pelicans. Walking a shell-strewn white Gulf beach in the early morning can be productive, as the roosting birds have not been pushed off the shore by all the human recreational activities that arise during the day.

During an early morning walk on the Gulf shore on Captiva Island on December 31, 2021, I saw 45 Red Knots, 40 Sanderlings, 10 Dunlins, 8 Ruddy Turnstones, and 1 Short-billed Dowitcher (as well as 20 Black-bellied Plovers, 1 Wilson's Plover, and 2 Killdeer). These birds were roosting on the beach, which itself was not a particularly productive foraging environment (except for the beach-loving Sanderlings). As the morning progresses and human activity picks up, these beach-roosting shorebirds depart for the bay side of the islands and the abundant mudflats to be found at low tide between the barrier islands and the mainland.

For a birder, the best access to the bayside mudflats is via kayak out from the eastern shore of Captiva or via bicycle down to J. N.

"Ding" Darling National Wildlife Refuge on northern Sanibel Island. The shallow waters on either side of Buck Key, on the bay side of Captiva Island, often feature little sandbars that attract shorebirds and terns. But the mix of bayside habitats at Ding Darling Refuge offer much more habitat for foraging shorebirds, including freshwater impoundments, mangrove edges, and extensive open mudflats. The refuge's wildlife drive gives access to all these rich habitats and rarely disappoints. Although Ding Darling is most famous for its big waterbirds—Wood Storks, Roseate Spoonbills, various herons, American White Pelicans, and such, it is also a good location for shorebirds during the midwinter months. Regular visitors include Spotted Sandpipers, both yellowleg species, Willets, Marbled Godwits, Ruddy Turnstones, Red Knots, Sanderlings, Western and Least Sandpipers, Dunlins, and Short-billed Dowitchers. Less common but occasional species include Whimbrels, Long-billed Curlews, Stilt Sandpipers, Long-billed Dowitchers, and Wilson's Snipes. Florida gets quite a few laggard Hudsonian Godwits in autumn, but this species apparently has a biological imperative to spend the northern winter in South America, so it is absent from Florida at that season.

For those birds, like the Hudsonian Godwit, that winter largely in South America, individuals found in the United States in the winter are mainly or exclusively birds of the year—juveniles that hatched in the summer up north. These are the individuals most likely to get lost, as they have never migrated to South America. For them the default strategy is to settle in where there is decent weather and productive feeding habitat, wait the winter out, and perhaps head back north in spring. Such off-kilter singletons have probably gotten separated from their migratory flock of juveniles on one of their initial flights out of their natal territory in the Far North. Unable to link up with another passing godwit flock, these individuals likely join up with other shorebird species heading south and end up stopping where these others stop. The main disadvantages to wintering in

Florida rather than southern Argentina are twofold: short day length and relatively meager food stores. Florida in winter is not a great place to fatten up for the next breeding season. By contrast, southern Argentina features the long days of Southern Hemisphere summer as well as the foraging riches that those southern environments generate during the height of summer.

Most shorebirds arrive on their Southern Hemisphere wintering grounds between August and November. Adults arrive exhausted and in poor condition. This is the time for them to recuperate from the extreme stresses of reproduction and migration. At this time, they carry out the postbreeding molt of their flight and contour feathers—a complete molt. Between November and February, the birds are on "holiday," loafing and taking it easy. Come late February, these birds need to get back to work: they molt a second time to produce their colorful breeding plumage, and they begin to sequester stores of nutrients to prepare them physically for the northbound migration and upcoming breeding season. These birds will start their journey northward in March. From August to February, as much as seven months, the birds are on the wintering ground. This is where they spend most of the year.

Molt

Birds molt annually because their feathers suffer a lot of wear and tear from all aspects of life through the seasons. Most shorebirds carry out two molts in a year: a complete molt after breeding, when they replace every feather on their body, and then a partial molt in spring that replaces their head and body contour feathering—generating the bright breeding plumage.

In general, molts take place at an energetically appropriate time, when demands are fewer. Molting is typically tucked in between the three biggest events in a bird's life: spring migration, breeding, and autumn migration. Molting takes a large amount of energy and typically occurs at a location rich in food resources and at a time when the bird is not facing other major life challenges (such as migration or breeding). Long-distance migrants tend to do a molt on the wintering ground, prior to migration. Birds that migrate shorter distances either do their molting on the breeding ground or during migration at staging sites where they rest and feed heavily over a period of several weeks. Some species start their molt on the breeding grounds in summer and autumn, continue it on migration, and complete it at the wintering grounds.

Molt of the flight feathers, called the primaries (the outer longer wing feathers) and secondaries (the inner wing feathers), is an important step in the annual molting cycle. Birds molt from the primaries closest to the body outward, and from the secondaries farthest from the body inward. During this process, there can be visible gaps in the wings in flight. In some species, the inner primaries are molted on the breeding ground and the outer primaries are molted on the wintering ground, with a hiatus of molt during migration to allow the birds to travel with a complete set of primary feathers.

The first year of a shorebird's life has three plumages: juvenile, winter, and first summer. After this first calendar year, most shorebirds exhibit only breeding and nonbreeding plumages. The first set of feathers grown after fledging is the juvenile plumage. This tends to be more heavily patterned on the back and scaley-looking as well. Often in flocks of a species of shorebirds in late summer, the juveniles' stronger dorsal pattern stands out while adults in autumn plumage look washed out. After the juveniles molt for the first winter, they will look more like the winter adults.

The adults of most species depart the breeding grounds in summer in their breeding plumage, molting either in transit or on the wintering grounds. First-year birds on the wintering grounds molt into a first-year breeding plumage, which in many cases is duller than the breeding plumage of older adults. Bar-tailed Godwits sometimes carry out a second molt of contour (body) feathering while staging in Europe in spring. The individuals that molt a second time are brighter and heavier than other birds, presumably giving them an advantage on the breeding ground. Unique among shorebirds, the Bristle-thighed Curlew carries out a complete molt in winter on South Pacific islands; this leaves the molting bird entirely flightless for a time.

Worn feathers are browner, frayed, and more pointed at the tip. Since the darker parts of feathers are more resistant to wear than the paler parts, wear is uneven, and birds with worn feathers become darker looking, especially when white feather tips wear off, leaving just the darker part. This, of course, changes the overall look of the bird. Molt counters these inevitable changes by bringing in fresh and bright feathering.

Wintering in the Pampas

Three of our North American shorebirds—the Upland Sandpiper, Buff-breasted Sandpiper, and Pectoral Sandpiper—winter primarily in the Pampas of South America, a large expanse of open and relatively aseasonal grasslands in the southern temperate zone. Encompassing 460,000 square miles in eastern Argentina, Uruguay, and southern Brazil, the Pampas are vast lowland plains that receive moderate annual rainfall—and which historically burn during the dry season. The burning prevents woody growth from overtaking these grasslands. The now-extinct Eskimo Curlew was also a member of the select shorebird group that favored wintering on interior grasslands rather than coastal shores. Such grasspipers look for grazed pastures with short grass where they can see the ground and spot their arthropod prey. These birds forage secondarily in both active agricultural and abandoned fields; rice fields are used on occasion. In the afternoon, these birds can be found visiting watering sites. Upland Sandpipers are sometimes found in taller grass fields and natural grasslands with light grazing impacts. Pectoral Sandpipers also visit marshy grounds and wetlands; this species has also been found in alpine-zone puna grasslands in Bolivia. The Pectoral is found in association with Buff-breasted Sandpipers and American Golden-Plovers.

Wintering in Southern Argentina and Chile

Several of our North American shorebirds spend the winter in southern South America, mainly in Argentina and Chile—Sanderlings, Red Knots, Whimbrels, White-rumped and Baird's Sandpipers, and Hudsonian Godwits. Unlike the preceding group, these shorebirds winter mainly in coastal habitats with ready access to tidal mudflats and other shoreline types. On the wintering grounds, Pectoral Sandpipers have been recorded feeding mainly on crustaceans, arthropods, and other invertebrates. Wintering Sanderlings inhabit sandy beaches just as they do in Florida and Delaware; Sanderlings wintering in Chile have been found to consume crustaceans, worms, mollusks, and arthropods. Wintering Red Knots in Brazil and Argentina roost on sandy beaches (as they do in Florida) and forage in intertidal flats (mud or sand); their winter diet is mainly small mollusks, especially bivalves such as mussels and clams. Whimbrels wintering in Argentina and Chile stay near the coast, foraging on tidal mudflats, coastal mudbanks, and to a lesser extent in marshes and mangroves; Whimbrels wintering in Venezuela have been seen feeding on fiddler crabs, as they do in Massachusetts on migration.

The Hudsonian Godwit winters from Paraguay south to Tierra del Fuego on the Atlantic side of South America and in southern Chile (especially Chiloé Island) on the Pacific. Chiloé Island is the wintering site for many or most of the Hudsonian Godwits that breed in Alaska and northwestern Canada—estimated to be 20,000 individuals in 2008. Winter roosting sites are of particular importance because flocks of thousands of shorebirds depend on relatively restricted areas where they can safely rest during high tide.

Wintering birds favor areas where there are secure roosting sites conveniently close to large bays that feature expanses of prey-rich coastal mudflats where the birds can forage at low tide.

Pectoral Sandpiper ***Calidris melanotos***
Other names: Grass-bird, Jacksnipe, Brown-back, Brownie, Marsh-plover, Creaker, Creaker Pert

Appearance: This large and trimly plumaged *Calidris* sandpiper has a crisply edged border separating its white underparts from its heavily streaked throat and breast. It has pale greenish legs.

Range: The Pectoral Sandpiper is one of the super-migrators, with some Siberian breeders wintering to Australasia. It summers in Arctic Canada, Alaska, and central and eastern Siberia. Autumn migration of the North American population passes mainly through the central sector of the Lower 48, and this population winters in central and southern South America.

Habitat: This sandpiper breeds in wet, grassy lowland tundra. Migrants favor grassy or muddy freshwater shallows.

Diet: The Pectoral Sandpiper's diet is mainly insects but also some crustaceans and plant matter. It combines pecking with rapid shallow probing, and it prefers foraging in grassy verges rather than in the water.

Behavior: This sandpiper is usually seen in small flocks. Its voice is a low-pitched *churk.* The male is polygamous and defends a large territory upon which multiple females nest. On the nesting ground, the male greatly expands his breast sac and conducts a striking breeding display for the female consisting of a low, slow flight with the chest distended, accented by a booming series of notes: *ooah ooah ooah ooah* for 10–15 seconds.

Nesting: The nest, set on a hill in grassy tundra, is a shallow depression with a lining of plant bits. The female lays four eggs that are drab, sometimes with a greenish tinge, and blotched with umber brown. Only the female incubates the eggs.

Conservation: The Pectoral Sandpiper is included on the NABCI Watch List. The species has declined as much as 65 percent since 1980. A 2012 population estimate was 1.6 million.

Hudsonians in winter use a range of additional coastal habitats: saltmarsh, freshwater and brackish lagoons, swamps, freshwater marshes, and flooded fields. Wintering birds feed primarily on marine worms and mollusks, which they harvest by probing mud. Researchers Morrison and Ross, in 1982–86, flew the coastline of South America censusing shorebirds. They counted 2.9 million shorebirds, and they located the main wintering grounds of the Hudsonian Godwit in Tierra del Fuego and southern Chile, where 45,000 godwits were tallied. Individual winter roosts of as many as 5,000 individuals have been recorded. The security of these large wintering godwit flocks is a high priority for conservation.

Wintering in the South Pacific

Three of our rarest North American shorebird breeders—the Wandering Tattler, Bristle-thighed Curlew, and Bar-tailed Godwit—winter in the tropical and subtropical South Pacific.

Nonbreeding-plumaged Stilt Sandpipers stop over in the Amazon, with (*left to right*) Common Gallinule, Black Skimmers, and Southern Lapwings in the background.

The tattler, which breeds in eastern Siberia, Alaska, and northwestern Canada, migrates down the Pacific coast of North America into Mexico; other birds fly across the Pacific to Hawaii, the southwestern Pacific Islands, New Zealand, and Australia. Birds wintering in the southwest Pacific are solitary inhabitants of rocky shorelines and offshore reefs, where they pick among coral rubble to harvest various marine invertebrates. Wintering birds remain in the Southern Hemisphere from August to February or March—as long as eight months.

The tiny population of Bristle-thighed Curlews departs its Alaskan breeding ground in July and wings south out across the Pacific. Some birds stop in the Hawaiian Islands; others continue nonstop to island groups in the South Pacific (Santa Cruz, Fiji, Tokelau, Tuvalu, Tonga, Samoa, Marquesas, Tuamotu, and Pitcairn Islands). Wintering birds forage on beaches, reefs, saltpans, lagoons, and short-grass habitats, such as the verges of airport runways. The species forages for crabs, mollusks, arthropods, and even birds' eggs, sometimes broken open with a stone. Subadult curlews will spend the northern summer on many of these Pacific islands, not returning to the Alaskan breeding habitat until a subsequent year.

Wintering on the Open Ocean

Two of our shorebirds, the Arctic-breeding Red Phalarope and the sub-Arctic-breeding Red-necked Phalarope, winter at sea. In fact, these are the only shorebirds on Earth to winter at sea, spending as much as nine months a year bobbing on the surface of the open ocean. These birds forage and roost in flocks at productive upwelling sites, thermal fronts, or in association with *Sargassum* seaweed mats, rich in nutrients and prey.

The Red-necked Phalarope winters primarily off the western coasts of Mexico, Central America, and South America, in the Arabian Sea south of the Arabian Peninsula, and in the tropical waters north of New Guinea and east of Borneo. The wintering Red-necked feeds on various small planktonic animal and plant life, especially copepods and euphausiids (two distinctive marine invertebrates).

The Red Phalarope winters off the western coasts of North, Central, and South America, off the coast of the southeastern United States and in the Caribbean, and in the eastern Atlantic off the coasts of West Africa and South Africa. The species is found in flocks foraging at offshore oceanic fronts where zooplankton are concentrated. The Red tends to winter farther offshore than the Red-necked and may consume prey that is smaller in size. Birds collected from offshore flocks had stomach contents that included copepods, amphipods, fish eggs, larval fish, and euphausiids. The wintering habits of these two phalaropes constitute the most extreme nonbreeding lifestyle of any shorebird.

Departing North from the Wintering Grounds

The path of the sun across the sky, in concert with the bird's endogenous biological clock, sends precise signals telling birds on their wintering grounds when they should depart northward. Males tend to leave first, as they are in a race to arrive on the breeding grounds early to take ownership of a productive nesting territory. In preparation for the flight north, the birds must molt into their breeding plumage and add fat and muscle to power the upcoming flights. This takes several months. It takes several weeks for the entire population to clear out, but known individuals have been shown to depart on the same date over multiple years. This was demonstrated in a color-marked and closely monitored New Zealand–wintering population of Bar-tailed Godwits. That said, there is plenty of flexibility shown by the birds. The presence of favorable winds may influence departure date, but each bird's endogenous clock appears to be the main determinant of departure time.

Heading northward, the larger species will make a few long flights, whereas the smaller species make a greater number of shorter jumps northward. At each stopping point, the birds will feed

The Phalaropes
1. Wilson's Phalarope ***Phalaropus tricolor***
Other names: Tri-color Phalarope, American Phalarope
2. Red-necked Phalarope ***Phalaropus lobatus***
Other names: Fairy Duck, White Bank-bird, Northern Phalarope
3. Red Phalarope ***Phalaropus fulicarius***
Other names: Whale-bird, Gray Bank-bird, Gray Phalarope, Herring-bird, Jersey-goose, Sea-goose, Sea Snipe

These unusual shorebirds can be broken into the ocean-going pair (Red and Red-necked) and the interior-loving singleton (Wilson's).

Appearance: The Wilson's in its winter plumage can be difficult to identify, looking a bit like a winter-plumaged Stilt Sandpiper. Notably, the breeding females are larger and more brightly colored.

Range: The Red and Red-necked breed across the Arctic and sub-Arctic zones and winter on the pelagic Atlantic, Pacific, and Arabian Sea. The Wilson's breeds in interior North America and winters to southern South America.

Habitat: The Red-necked and Red breed on tundra. The Wilson's breeds in association with alkaline lakes or freshwater marshes of the Interior West and Midwest. The two oceanic species winter to oceanic waters. The Wilson's winters to high alpine lakes of the Andes of southern South America.

Diet: The phalaropes' diet is mainly aquatic and marine invertebrates and occasionally tiny fish, picked from the freshwater or ocean surface. These three species are the only shorebirds that twirl in the water.

Behavior: All three species are sometimes polyandrous, with the larger and more colorful females carrying out the courtship display. The males build the nest and attend to the nestlings.

Nesting: The nest is a shallow scrape lined with plant matter. The female lays three or four olive or buff eggs blotched with brown.

Conservation: The IUCN Red List treats these three species as Least Concern. All three have populations in excess of a million.

voraciously to replenish fat stores and build muscle. For the tundra breeders that winter in southern Argentina, this northward journey may take up much of March, April, and May.

Timing

Wintering birds need to time their migration to arrive at their northern breeding grounds at the proper moment in the spring. One might think this is just a matter of changing daylight regimes and the sun's position in the sky at noon, but consider the many places where the Sanderling winters. Sanderlings wintering in southern South America will be seeing ever shortening day length and the sun lowering in the sky. Those wintering near the equator will see little change in solar position or day length. Those wintering in Massachusetts will see the days lengthening and the sun rising ever higher in the sky at noon. How do Sanderlings, facing wildly different solar and astronomical conditions, manage to arrive at their Arctic tundra breeding ground at the proper moment? If they arrive too early, the ground will be frozen under a thick layer of snow. If they arrive too late, they will find all the best territories held by other males, with females already having chosen mates. Each Sanderling must have an internal clock that guides it northward at the correct moment. The details of this endogenous clock remain a biological mystery. Approaching the Arctic breeding zone, the birds need to be flexible, reading additional cues that prevent them from arriving on the tundra still fully carpeted in deep snow.

Summing up the "wintering" experience for shorebirds, there are several points worth making. First and foremost, the nonbreeding season is long and the different shorebird species exhibit great variability in their wintering movements and activities. Second, this tends to be the time when we interact with shorebirds with the greatest frequency. We are used to seeing the Sanderlings in their pale winter plumage. Recall also that the wintering and nonbreeding modality in the shorebirds is a safe default strategy that is employed in other seasons by young birds and birds that are unable to compete on the Arctic breeding grounds (when in doubt, hang out on the wintering grounds). Finally, the term "wintering" is, in many cases, a misnomer. The most successful super-migrators spend the various yearly seasons in search of spring and summer climes, with long day length and abundant food resources. Winter is avoided. Those long migratory flights make this successful strategy possible. ❋

12 The Magic of Epic Migration

In the days of our grandfathers the Gray-backs or Wahquoits, as they were called [the Red Knots], *swarmed along the coasts of Cape Cod by the thousand. Spring and autumn their hosts were marshaled on the flats of Barnstable County, and around Tuckernuck and Muskeget Islands they collected in immense numbers and rose in "clouds" before the sportsman's gun.*

—Edward Howe Forbush, *Birds of Massachusetts and Other New England States*

SCIENTISTS CONTINUE TO STRUGGLE to offer a credible and complete explanation for how our scolopacid shorebirds manage to migrate successfully, year after year, from the Northern Hemisphere to the Southern Hemisphere and back again. Because the migratory process is complex, let us look at the tools the birds depend upon, one by one. If we can explain its components, perhaps we can approach explaining the whole.

An Internal Clock

First and foremost, the migrating shorebird possesses a "clock." In fact, it is a clock with a series of annual alarm settings, marking important points in the shorebird's year. Here are two examples of the clock's specificity. A color-tagged Bar-tailed Godwit has departed northbound from New Zealand within a day or two of March 25 year after year over a 13-year period. And recall the color-marked birds arriving in James Bay on the exact anniversaries of their previous arrival dates. This shows that shorebirds can travel from point to point in a timely fashion across the face of the Earth following a fixed and exacting timetable.

Birds possess an internal biological clock, driven by hormones and presumably calibrated by signals received from the position of the sun in the sky, which lets them know the time of day as well as the time of the year. These clocks are recalibrated by external cues—photoperiod (the day length that changes with the seasons) being one of the most important—keeping them accurate from season to season and year to year. Long-term experiments have shown that these internal clocks can operate even in the absence of the external cues. Most remarkable is that the birds can quickly recalibrate their clock when crossing from the Northern to Southern Hemisphere, where photoperiods and day length are transposed. This internal clock is critical to the livelihood of these migrant birds, making certain they arrive on the breeding ground at the appropriate moment and that they head off on long overwater migrations when likely weather conditions are most favorable for such flights.

Critical Earth Locations

All North American shorebirds are migratory. What makes these journeys particularly remarkable is philopatry, the tendency of some animals to return, year after year, to specific sites important to their lives—the site of their birth, the site of first breeding, their main wintering site, and key staging sites during migration. In some cases, individual birds return to these sites year after year over a life of more than 20 years. These few critical sites are analogous to logged GPS points, perhaps with marked waypoints between them. In some instances, these sites may be inherited and hardwired in the bird's brain, specifically the hippocampus, where spatial information is stored. It has been shown that some birds grow new neurons to enlarge the hippocampus, which probably aids their navigation abilities. Much discussion of bird migration focuses on orientation and navigation. But why orient and navigate unless there is a fixed destination to travel to?

The most important single point on Earth to a godwit is its birthplace. All subsequent travels are carried out with a geographic reference to this site. One must assume that the godwit possesses an array of inborn Earth-positioning cues and abilities that enable it to return to this natal site. We can safely assume that the bird possesses a set of allied skills that act like one of our handheld GPS tools. Key in "Home," and the tool will guide the bird back to its birth site. As the bird travels out from its birth site, this tool adds new waypoints, one after another. As these accumulate, they generate an ever-richer internal map of the Earth for that migrant bird.

GPS Tools

We have long known that migrating birds use solar and celestial cues to aid their orientation on Earth. The path of the sun across the sky and the points on the horizon where the sun rises and sets provide vital information to the migrating bird about the cardinal directions. And the bird's ability to detect polarized light further extends the period in which it can see sunrise and sunset. The bird combines data from its internal clock with information about the sun's position to determine compass direction. After dark, migrating birds rely on the nightly rotation of bright stars and constellations around the polestar to orient themselves. The birds do not cue on constellations themselves but rather on stellar movement to establish compass direction.

Birds are further aided by an endogenous magnetic sense. In the late twentieth century, field and laboratory experiments demonstrated that Homing Pigeons could orient based on the magnetic fields of the Earth, even when they are unable to see the sun or the night sky. For decades, it was thought that tiny crystals of magnetite in the birds' foreheads provided the basis for an internal compass. But it turned out that birds were detecting magnetic field lines of the Earth, not the simple north–south directionality of a compass.

Current theory proposes that a pigment protein in the retina of birds, known as cryptochrome 1a, enables birds to detect the Earth's magnetic field lines; these bow out from each pole, aligning at the equator. Light in the blue wavelength allows birds to "see" these magnetic field lines. Deutschlander and colleagues conducted experiments that showed that bathing birds in red light actively prevented birds from orienting themselves using their magnetic sense. Only the blue wavelength provided this signal to the birds. Being able to "see" these magnetic field lines allows birds to detect magnetic inclination, which is horizontal at the equator but deeply tilted downward when near each pole. Thus, as the migrating bird moves toward one of the poles, the detection of the dip of the magnetic field lines gives the navigating bird an accurate measure of its latitude. The magnetic field lines also provide the migrating bird an exact measure of magnetic north and south.

The physics of the bird's deployment of cryptochrome 1a involves quantum mechanics beyond the understanding of most field biologists: it requires knowledge of radical pairs, quantum entanglement, spin, a photon, and "spooky action at a distance" (to quote Einstein) to understand a bird's magnetic compass. So forget about the magnets in the forehead; birds use quantum mechanics to orient themselves on the Earth.

Research by Wynn and collaborators on Eurasian Reed Warblers has demonstrated that as a migrating bird flies northward in spring, three aspects of Earth's geomagnetism come into play—inclination, declination, and magnetic field strength. In the first, the magnetic field lines tilt further downward as the bird continues northward. In the second, the magnetic field declines to the east or west relative to the true North Pole. The third ingredient, Earth's magnetic field strength, which is strongest at the North Pole and weakest at the equator, provides the bird with an additional tool for assessing its location.

Magnetic inclination and declination plus geomagnetic strength of the natal site are internalized by the yearling birds before they take off for the south for the first time. On the northbound return, inclination tells the migrating bird to stop at a particular latitude. It can then use the magnetic declination (deviation from the true pole) to determine whether to travel east or west to its target site.

Other cues are available. Some birds have an extraordinary sense of hearing that allows them to detect infrasound. Low-frequency sounds have the capacity to travel long distances. Such sounds are produced by Earth's topographic features—for example, winds striking a mountain ridge or an ocean's waves striking a shore. These low-frequency sounds also provide locational information to a migrating bird that has previously flown through that territory. Yet another cue may be visible wave patterns on the sea surface, seen from high elevation by the migrating bird. Visible wave patterns may combine with infrasound cues to provide critical geographic information to Bar-tailed Godwits traveling high over the sea.

Believe it or not, some birds have an excellent sense of smell, and they can potentially use scents to orient. (Think of the scent given off by a mature forest of Jack Pines in the Far North.) These abilities may all contribute to the bird's success in orienting itself on Earth.

One might assume that a migratory flock of adult godwits headed south over the water will have a dominant "leader" that is older and more experienced. Indeed, the flock may have several such birds. Following signals from these more experienced birds, the flock can presumably integrate years of experience with the orientation and navigation tools that all the flockmates possess.

Birds do not always take the shortest or most direct route to a destination. One group of Bar-tailed Godwits, migrating over water from the Yellow Sea to Alaska, had timed departures associated with favorable winds, but those winds were linked to a large-scale weather system that delivered them south of their target destination. At the last moment, these birds succeeded in making a series of substantial course corrections that brought them to their intended destination.

Physiological Adaptation

The navigational abilities of our super-migrators are jaw dropping. But that is not the only capacity that makes these birds special. The physiological adaptations the birds exhibit, and their ability to make these changes rapidly, defy the imagination.

Fat supplies provide the critical fuel that powers the super-migrators' days-long, overwater, continuous-flapping flights. Burning fat rather than other nutritional sources (sugars, proteins) allows these birds to achieve seemingly impossible feats of endurance. To make this possible, staging birds add as much as 100 percent of their body weight in stored fat in a matter of weeks. The liver may get two to three times bigger to be able to convert stored protein and sugars to fat. The absolute limitation on fat storage seems to be the ability to take flight; some birds (Sanderlings, Bar-tailed Godwits) have been observed to have difficulty getting off the ground because of their newly gained weight. The maximum fat load determines not the distance the bird can traverse but rather the duration of continuous-flapping flight. Maximum distance, then, is determined by the flapping flight duration multiplied by the mean ground speed of the bird over the length of the flight.

Paralleling weight gain, the birds pack on a large mass of new pectoral muscle to power their wing beats for the long-distance flight. Bar-tailed Godwits double or triple the mass of their pectoral muscles and increase the mass of their heart, their blood cell concentrations, and their lungs before a big flight. One interesting study by Dietz and colleagues found that captive and constrained Red Knots grew breast muscle just prior to their annual migration departure date, suggesting that an endogenous calendar drives the physiology of these birds.

As the day of departure approaches, the bird rapidly restructures its internal organs. It shrinks its gizzard, liver, gut, and kidneys to reduce its organ weight to offset the countervailing gain in pectoral muscle and fat stores. In essence, the bird shuts down and minimizes its feeding system for its long migratory flight.

Godwits have been tracked flying long distances at high altitudes—as high as 20,000 feet above sea level. Presumably, the hypertrophied lungs of these birds allow them to extract sufficient oxygen and avoid hypoxia at these altitudes while burning their belly fat with never-yet-measured levels of efficiency. Physiological modeling predicts the energy stores of these birds should give out in 3–4 days, not the 9–11 days that the birds are known to accomplish. The current record for sustained flapping flight by a Bar-tailed Godwit is 265 hours, while operating at eight times its base metabolism. The migrating bird achieves this through rates of body mass loss and energy expenditure that are less than half the rates predicted by a half-century of studies of the physiology of avian flight; this is to say the migrating bird manages to achieve the impossible. Physiologists need to go back to the drawing board to explain this phenomenon.

Migrating godwits are unable to drink for the length of their multiday flight. But birds can address the issue of water loss and hydration by directly burning fat to generate wing stroke power. The burning of fat generates CO_2 and water, and this recovered water can apparently be deployed to meet the bird's physiological demands during migration.

How do these godwits cope with going without sleep for 11 days? This is another challenge needing a physiological explanation. It has been hypothesized that some birds can shut down half their brain and let that side of the brain sleep while the other side remains awake. Does this concept of unihemispheric sleep apply to shorebirds on migration? We do not know. We know that Alpine Swifts wintering in Africa keep to the air without alighting for more than 200 days, apparently practicing unihemispheric sleep. Is sleep indeed possible during extended flapping migratory flight over days? Again, that is not known. Godwits arriving in New Zealand after a long flight often sleep for a substantial period even before they resume feeding, despite their long fast. Rather than unihemispheric sleep, these birds may experience "adaptive sleep loss." More research on this front is needed.

Weather Detection

But what of weather forecasting by the birds? Migrating birds must potentially contend with cloud, fog, rain, air temperature, approaching storms, and of course wind speed and direction (which varies with altitude). Most birds initiate a migratory flight only when weather conditions are favorable—clear skies, light winds, and wind direction pointing toward the bird's destination.

Hudsonian Godwits, traveling for several days nonstop over the Atlantic, can use "adaptive wind drift" to make a substantial course correction when the route is threatened by an unanticipated major wind system. Such skills require all sorts of capabilities that seem to be well beyond the capacity of an adult bird with a brain the size of a hazelnut.

Apparently, Bar-tailed Godwits can divine favorable wind and weather for their stupendous flights. In 2014, Gill and colleagues noted that "this behavior suggests that there exists a cognitive mechanism, heretofore unknown among migratory birds, that allows godwits to assess changes in weather conditions that are linked (i.e., teleconnected) across widely separated atmospheric conditions." This is not explained by current biology. But scientists have tracked airborne flocks making major shifts in altitude to catch beneficial winds during a long-distance flight. In other words, these migrants appear able to "see" distant atmospheric conditions and make flight changes to take advantage of them. Interannual variation in the wind systems (such as El Niño) adds to the complexity of this challenge.

And how will climate change's influence on major weather systems impact the success of trans-Pacific migrants in decades to come?

First Southbound Migration

It is difficult to imagine how groups of 60-day-old Bar-tailed Godwits gather on the breeding ground in western and northern Alaska and, before long, without any guidance or attendance from adult Bar-tailed Godwits, travel to a staging site in coastal Alaska where they have never been, fatten up, and then launch off on a long overwater flight to a destination in the Southern Hemisphere they have never seen. Juveniles accomplish the same flight from Alaska to New Zealand as the adults. How is this possible?

The only adequate explanation is that many cues and travel rules are hardwired, inherited from the successful super-migrators of preceding generations. Learning can also be part of the process, as has been demonstrated by the clever translocation experiment with juvenile Black-tailed Godwits in Europe (see page 132). But presumably learning is subordinate to the hardwired rules that guide that first flight south. These birds with mind-blowing navigational capacities are driven by DNA-coded instructions based on extremely rigorous regimes of natural selection, where only those who survive a southbound and subsequent northbound flight will pass on their genes. We can be confident that the hardwiring has been exquisitely fine-tuned by stiff selection gradients driven by the crucible of the globe-spanning trans-Pacific migrations.

The Super-Migrator Strategy

Studies have shown that long-distance migrants have longer staging stopovers, add more weight and fat, and fly at higher altitudes than short- and medium-distance migrants. Those with the greatest fuel loads are the least likely to stop over in the Lower 48. In sum, super-migrators seek to minimize travel times and reduce rest stops that make birds vulnerable to predation and other mishaps. Adult super-migrators live longer than smaller and more typical shorebirds, supporting the notion that these birds have evolved super-long migrations to provide maximum opportunity for survival. It may sound counterintuitive, but those shorebird species that carry out the most extreme migrations appear to be the species with the lowest adult mortality rates. The long flights probably generate higher mortality in juveniles, but substantially lower mortality for the older, experienced birds. These successful adults live 10 to 20 years and spend most of their long lives in perpetual spring with long days and ready access to abundant food resources. By avoiding cold winters and stopover sites infested with hungry predators, these birds live the good life, and for quite a long time. They manage to do this because their species evolved to carry out super-migrations that depend upon remarkable navigational abilities that remain only partially explained. They move easily across the face of the Earth with a powerful internal GPS capacity, a product of the harsh regimes of evolution. Let's tip our hat to this wonder of nature!

There is plenty of science yet to be done to properly and fully explain the migratory abilities of these shorebird super-migrators. We need to better understand the workings of the shorebird's magnetic sense and its operation through quantum entanglement. And how does this special sense translate into a GPS-like navigational ability by the bird when moving across the Earth? Most puzzling of all is the ability of yearling birds, born in Alaska, to migrate to Southern Hemisphere destinations successfully without prior experience or the assistance of adult birds. How can an internal map complex enough to guide these yearlings from Alaska to New Zealand be hardwired in the bird's tiny brain? And what about the physiology of 11-day-long non-stop flight? The chemistry of superefficient use of stored fat needs to be upgraded to explain this feat. What about the long-distance detection of Pacific-wide weather systems to benefit from winds at certain altitudes and to avoid storms? Godwits can do that, but we do not know how. Indeed this research awaits.

Are there new tools that could be deployed to address these biological mysteries? Certainly lightweight satellite-tracked transmitters can add new sensing capacities to measure a bird's physical condition and movement. Following the birds in real time should allow us to see how they respond to changing atmospheric conditions and weather challenges. One might expect a combination of laboratory-based and field experimentation could begin to crack the code on these other migration mysteries. And, no doubt, artificial intelligence will be tapped to help address all of the computing demands that will be needed to crunch the very large datasets collected on the bird's physiology and the weather conditions the migrating bird is experiencing and processing. At this point, we simply must admire the stupendous accomplishments of our epic migrators. ✵

Epilogue

The beauty and genius of a work of art may be reconceived, though its first material expression be destroyed; a vanished harmony may yet again inspire the composer; but when the last individual of a race of living beings breathes no more, another heaven and another earth must pass before such a one can be again.

—C. William Beebe, *The Bird: Its Form and Function*

My Magnificent Seven led me far and wide across North America, from coast to headwaters and from prairie to tundra. Tracking down these sublime shorebirds took me out on the lonely road and landed me in beautiful landscapes little known to most of us. Searching for rare migrants is an adventure that spawns new memories of special places—a bird in a place at a particular time. It's a Whimbrel in Maine in 1969 or a Bristle-thighed Curlew in Alaska in 2022. The colors, the sounds, the scents, and the sense of it all weave a tapestry of experience that halts time. It is the permanence of a treasured memory that we are hunting in nature that raises us above the monotony of the day-to-day. That is our elusive objective.

Driving the back roads, solo, searching, we find America's most special green spaces. These are not the crowded vistas near a filled parking lot. They are stumbled upon, by chance, by mistake, without a plan. An Alaska breakfast under deep blue skies with the snowy dome of 16,000-foot Mount Sanford dominating the eastern horizon. A glimpse of Tombstone Peak at the head of a glacier-scoured valley. Or multicolored badlands where a young Theodore Roosevelt once hunted bison. So many special places to see for the first time!

And abundance! Seeing wildlife we love in abundance fills our hearts with joy. By visiting certain places during the migration season, we can see 100 Whimbrels, or a handsome cluster of godwits, or a mudflat swarming with Semipalmated Sandpipers. But we read reports of shorebird decline. Serious decline. Even the commonplace species are declining. That news of decline makes our hearts ache. On the one hand, we want to celebrate the abundance we see, but then again, we acknowledge the need to battle against the species' declines. That's the intense bittersweet of today.

Shorebirds suffered greatly in North America during the market-hunting era of the late nineteenth and early twentieth centuries. Shooting of the Eskimo Curlew in New England in autumn and in the Mississippi drainage in spring led to its near extinction by the 1910s. The last Eskimo Curlew collected on Cape Cod was a singleton taken at East Orleans in 1913. Roland Clement observed a single bird at Race Point atop Cape Cod in 1922. Migrating curlews also faced habitat loss from agricultural conversion of prairies in the Great Plains, as well as from the mysterious disappearance of the Rocky Mountain Grasshopper, a prime food for the northbound migrant each spring. In his *Handbook of Birds of Eastern North America,* Frank Chapman reported this curlew to be "nearly or quite extinct" in 1939.

Upland and Buff-breasted Sandpipers, the Whimbrel, the Long-billed Curlew, and the Hudsonian and Marbled Godwits all were heavily harvested for the table and the market. Roosting Red Knots were collected at night by hand when hunters temporarily blinded flocks of them using fire torches. Writing in the ornithological journal, *The Auk*, in 1897, regarding the overhunting of shorebirds, George H. Mackay asked querulously, “Are we not approaching the beginning of the end?” None of the shorebirds listed above has returned to its original abundance, even though all but the Eskimo Curlew have managed to survive into the current century.

Public reaction to the mass slaughter of birds for the table, and for the plumes that went to decorate women’s hats, prompted the formation of various local and national bird conservation organizations in the United States. These organizations successfully lobbied for the Migratory Bird Treaty Act, signed with Canada in 1916 and with Mexico in 1936. Although the first federal law controlling the spring hunt of migratory shorebirds was passed in 1913 (the Weeks-McLean Law), fall hunting of shorebirds continued until 1927—a grim time when hunters had nothing to shoot at but tiny peeps and Tree Swallows because their favorite targets had been decimated.

After market hunting was halted, most shorebird populations recovered to some extent. However, population losses continue. Habitat alteration in wintering and migratory stopover areas is one culprit. Market, subsistence, and sport hunters in the Caribbean and in Central and South America continue to shoot migrating shorebirds. In many instances this hunt is legal. Climate change may become the most serious threat to these birds. As the permafrost that underlies Arctic tundra melts, it will certainly alter the birds’ breeding habitats and the availability of their summer prey base. Growing mismatches between spring climates and the breeders’ arrival times may become a threat. The birds are adapting, but the question is whether they can adapt fast enough to keep pace with the climate shift in the Far North, where temperature increases are the most extreme on Earth.

Overall, North American shorebirds are in decline, some substantially. Those migrating along the shores of East Asia and the Yellow Sea are particularly impacted by the building of seawalls, pollution, degradation of critical staging wetlands, illicit bird-netting, fishnet bycatch, and commercial aquaculture development. For many Arctic-breeding trans-Pacific migrants, the Yellow Sea hotspot is a critical conservation battleground.

In a 2013 multi-institutional study of 28 species of North American shorebirds, Smith and colleagues found that more than half were in substantial decline. Some of the species most affected were the Red Knot, Hudsonian Godwit, and Short-billed Dowitcher. Most alarming, the rates of decline appear to be increasing. (The study found comparable levels of decline only in the North American aerial insectivores and grassland birds.) Similar declines have been documented among shorebird populations in Australia, presumably because of similar poorly understood impacts. What exactly is causing these wholesale declines is unclear, but it is probably an array of impacts working in concert. Most ominously, the substantial decline of shorebirds as a group has the look of the beginning of a global extinction event.

This news should be setting off alarm bells in the conservation departments of federal and state governments as well as within the leadership teams of national and international conservation organizations. Our beloved shorebirds are under serious threat. Studies suggest that even though most adult super-migrators survive their long-distance travels, this lifestyle is precarious at the population level, and it could lead to population crashes when environmental conditions change. Even the familiar Sanderling has declined by about 40 percent since 1980. Is it too late to turn things around?

The story of the Atlantic coast populations of the American Oystercatcher (a shorebird relative) suggests that it is perhaps not too late. When this shore-loving species was shown to be in serious decline in the early 2000s, a coordinated multi-institutional and multistate program of action was implemented, and today the Atlantic populations have rebounded substantially. Let's tip our hats to some of the institutions that made this happen: Manomet Center for Conservation Sciences, US Fish & Wildlife Service, US Geological Survey, the Nature Conservancy, New Jersey Audubon, National Audubon Society, Massachusetts Audubon, National Fish and Wildlife Foundation, and the Virginia Department of Wildlife Resources.

A similar story can be reported for the American Woodcock. Habitat improvement programs for this beloved species have been carried out by state governments working in partnership with the American Bird Conservancy across the woodcock's breeding habitat. These local habitat interventions have immediately benefited breeding woodcock populations.

Working with ranchers on private lands in the arid West, the American Bird Conservancy (ABC) has taken on conservation of breeding populations of the Long-billed Curlew, and to good effect. These are now being expanded to additional breeding locations and to wintering grounds in Texas and Mexico to multiply the beneficial impacts.

The take-home point here is simple: conservation actions, when well designed, properly funded, and carried out with full cooperation between the necessary institutions and communities, do indeed work. The task at hand is, first, to divine the nature of the threats to our shorebirds and then to bring the players together to construct smart interventions. These need to address the threats to the birds on their breeding, stopover, and wintering grounds. That's a big job!

And let's not overlook another important international program—the Western Hemisphere Shorebird Reserve Network. WHSRN is a partnership based in science that protects critical habitats for shorebirds throughout the Americas. It currently fosters the conservation of 124 special shorebird habitats from Alaska to southern Argentina. These are staging and wintering sites where shorebirds congregate in large numbers—sites such as the Delaware Bay shore of New Jersey and the Copper River Delta in Alaska. These are the jewels in the crown of the Western Hemisphere's most precious shorebird reserves. Supporting ABC and WHSRN will help address the threat of shorebird decline.

Let's all put our shoulder to the wheel. Every reader of this book should invest in local, national, and international conservation organizations that engage in shorebird conservation—institutions such as ABC and WHSRN. We all want future generations to have an opportunity to glimpse a Whimbrel on a rocky Maine shoreline or to scope a Hudsonian Godwit on a Cape Cod mudflat before it heads south out across the Atlantic, bound for the productive south temperate spring climes of Chile and Argentina. ❋

Acknowledgments

I OFFER SINCERE THANKS to all those who made possible the various solo field trips that underpin this book. Certainly that starts with my wife, Carol, who kept the home fires burning while I was traipsing about places as diverse as Alaska, Manitoba, the Yukon, North Dakota, and even South Texas.

Many generous friends underwrote the costs of carrying out the field trips, preparing the art, and editing and designing this book. They include Jane Alexander, Lisa Anderson, Liz Dugan, John Friede, Caroline Gabel, Rampa Hormel, Jared Keyes, Warren King, Maria Semple and George Meyer, Hugh Rienhoff, Carol Sisler, and John Swift. Thank you very much!

My base of ornithological studies remains the Bird Division of the National Museum of Natural History (Smithsonian Institution). I thank Helen James, Gary Graves, Chris Milensky, Brian Schmidt, Christina Gebhard, Joe Jehl, and Carla Dove for their support and collegiality.

The American Bird Conservancy (ABC) served as institutional sponsor for the godwit project, and as part of this relationship, I serve as a scientific affiliate of ABC. I thank Mike Parr, Erin Chen, Jack Morrison, and Clare Nielsen for their collaboration, guidance, and assistance.

Fellow bird people, with whom I spent happy times watching godwits and other shorebirds, include Michael O'Brien and Louise Zemaitis, David Wilcove, Christian Caryl, Meg and John Symington, Patsy and Tom Inglet, and Jane Tillman.

Nathan Senner and Maria Stager hosted me at the godwit breeding bogs in Beluga, Alaska. It was great to participate with them as they did their research. While I was in Alaska, I was assisted in various ways by William Fredette, Max Vigeant, and David and Andy Sonneborn. Julie Hagelin and Peter Elsner welcomed me to Fairbanks, Alaska. LeeAnn Fishback hosted me at the Northern Studies Centre in Churchill, Manitoba. Bob and John Romohr showed me the ropes around Gresham, Nebraska. Christian Friis of the Canadian Wildlife Service made it possible for me to participate in the autumn shorebird surveys in James Bay. While I was at Longridge Point, Doug McCrea and Ross Wood were wonderful guides and hosts.

Artist Alan Messer devoted months to crafting the beautiful illustrations of shorebirds for this work. Working with Alan on this project was a gift. John Anderton painted the flying godwits that grace the cover. Bill Nelson prepared the maps that help tell the story of the book. Carrie Love meticulously copyedited the book. And Jaime Schwender, Jennifer Linscott, Judith Schindler, John Kricher, and David Wiedenfeld provided critical reads of the book, making hundreds of improvements. I thank Carolyn Gleason and Smithsonian Books for agreeing to take on this project. Chuck Burg and Lela Stanley reviewed the page proofs, catching errors many of us missed at earlier stages. Finally, I salute Carol Beehler, graphic designer, for her eye for beauty.

Bibliography

Anderson, Alexandra M., Sjoerd Duijns, Paul A. Smith, Christian Friis, and Erica Nol. 2019. "Migration Distance and Body Condition Influence Shorebird Migration Strategies and Stopover Decisions During Southbound Migration." Frontiers in Ecology and Evolution 7 (251): 1–11.

Andres, Brad A., James A. Johnson, Jorge Valenzuela, R. I. Guy Morrison, Luis Espinosa, and R. Ken Ross. 2009. "Estimating Eastern Pacific Coast Populations of Whimbrels and Hudsonian Godwits, with an Emphasis on Chiloé Island, Chile." Waterbirds 32 (2): 216–224.

Audubon, John James. 1835. *Ornithological Biography*. Vol. 3. Edinburgh: Adam and Charles Black.

Beebe, C. William. 1906. *The Bird: Its Form and Function*. New York: Henry Holt.

Beehler, Bruce M. 2018. *North on the Wing: Travels with the Songbird Migration of Spring*. Washington, DC: Smithsonian Books.

Bent, Arthur Cleveland. 1962. *Life Histories of North American Shorebirds*. Parts 1 and 2. New York: Dover Publications.

Černý, David, and Rossy Natale. 2022. "Comprehensive Taxon Sampling and Vetted Fossils Help Clarify the Time Tree of Shorebirds (Aves Charadriiformes)." *Molecular Phylogenetics and Evolution* 177:107620.

Chapman, Frank M. 1966. *Handbook of Birds of Eastern North America*. New York: Dover Publications.

Conklin, Jesse R., and Phil F. Battley. 2011. "Impacts of Wind on Individual Migration Schedules of New Zealand Bar-Tailed Godwits." *Behavioral Ecology* 22 (4): 854–861.

Conklin, Jesse R., Nathan R. Senner, Philip F. Battley, and Theunis Piersma. 2017. "Extreme Migration and the Individual Quality Spectrum." *Journal of Avian Biology* 48:19–36.

Cramer, Deborah. 2015. *The Narrow Edge: A Tiny Bird, an Ancient Crab, and an Epic Journey.* New Haven, CT: Yale University Press.

del Hoyo, Josep, Andrew Eliott, and Jordi Sargatal, eds. 1996. *Handbook of the Birds of the World*. Vol. 3, *Hoatzin to Auks*. Barcelona: Lynx Edicions.

Deutschlander, M. E., J. B. Phillips, and S. C. Borland. 1999. "The Case for Light-Dependent Magnetic Orientation in Animals." *Journal of Experimental Biology* 202:891–908.

Dietz, Maurine W., Theunis Piersma, and Anne Dekinga. 1999. "Body-Building Without Power Training: Endogenously Regulated Pectoral Muscle Hypertrophy in Confined Shorebirds." *Journal of Experimental Biology* 202:2831–2837.

Dunn, John L., and Jonathan Alderfer. 2017. *National Geographic Field Guide to the Birds of North America*. 7th ed. Washington, DC: National Geographic Society.

Egevang, Carsten, Iain J. Stenhouse, Richard A. Phillips, Aevar Petersen, James W. Fox, and Janet R. D. Silk. 2010. "Tracking of Arctic Terns *Sterna paradisaea* Reveals Longest Animal Migration." *Proceedings of the National Academy of Sciences* 107:2078–2081.

Emlen, Stephen T. 1970. "Celestial Rotation: Its Importance in the Development of Migratory Orientation." *Science* 170:1198–1201.

Forbush, Edward Howe. 1925–1929. *Birds of Massachusetts and Other New England States*. 3 vols. Boston: Commonwealth of Massachusetts.

Forbush, Edward Howe, and John Bichard May. 1955. *A Natural History of American Birds of Eastern and Central North America*. New York: Bramhall House.

Gauthreux, Sidney A. 1971. "A Radar and Direct Visual Study of Passerine Spring Migration in Southern Louisiana." *Auk* 88:343–365.

Gill, R. E., Jr., T. L. Tibbitts, P. F. Battley, N. Warnock, and D. C. Douglas. 2023. "Tracking Data for Bar-tailed Godwits (*Limosa lapponica*)." Ver 1.0, November 2023. US Geological Survey data release. https://doi.org/10.5066/P9A9BYQW.

Gill, Robert E., Jr., David C. Douglas, Colleen M. Handel, T. Lee Tibbitts, Gary Hufford, and Theunis Piersma. 2014. "Hemispheric-Scale Wind Selection Facilitates Bar-tailed Godwit Circum-Migration of the Pacific." *Animal Behaviour* 90:117–130.

Hagar, Joseph A. 1966. "Nesting of the Hudsonian Godwit at Churchill, Manitoba." *The Living Bird* 5:5–43.

Hall, Henry Marion. 1960. *A Gathering of Shorebirds.* New York: The Devin Adair Company.

Kaufman, Kenn. 1996. *Lives of North American Birds.* Boston: Houghton Mifflin Company.

Lehnen, Sarah E., and David G. Krementz. 2005. "Turnover Rates of Fall-Migrating Pectoral Sandpipers in the Lower Mississippi Alluvial Valley." *Journal of Wildlife Management* 69 (2): 671–680.

Liang, Dan, Tong Mu, Ziyou Yang, Xingli Giam, Yudi Wang, Jing Li, Shangxiao Cai, Xuelian Zhang, Yixiao Wang, Yang Liu, and David S. Wilcove. 2023. "Assessing Shorebird Mortalities Due to Razor Clam Aquaculture at Key Migratory Stopover Sites in Southeastern China." *Conservation Biology* 38:e14185.

Linscott, Jennifer A., and Nathan R. Senner. 2021. "Beyond Refueling: Investigating the Diversity of Functions of Migratory Stopover Events." *Ornithological Applications* 123:1–14.

Loonstra, A. H. Jelle, Mo A. Verhoeven, Christiaan Both, and Theunis Piersma. 2023. "Translocation of Shorebird Siblings Shows Intraspecific Variation in Migration Routes to Arise after Fledging." *Current Biology* 33 (12): 2535–2540.

Lovette, Irby J., and John W. Fitzpatrick. 2016. *The Cornell Lab of Ornithology Handbook of Bird Biology.* 3rd ed. Hoboken, NJ: Cornell Lab of Ornithology and John Wiley and Sons.

Lowery, George H. 1945. "Trans-Gulf Spring Migration of Birds and the Coastal Hiatus." *Wilson Bulletin* 37:92–120.

Mackay, George A. 1897. "The 1896 Migration of *Charadrius dominicus* and *Numenius borealis* in Massachusetts." *Auk* 14:212–214.

Matthiessen, Peter. 1973. *The Wind Birds*. New York: Viking.

Mayr, Ernst. 1953. "On the Origin of Bird Migration in the Pacific." *Proceedings of the 7th Pacific Science Congress* 4:387–394.

Morrison, R. I. Guy, and Richard Kenyon Ross. 1989. *Atlas of Nearctic Shorebirds on the Coast of South America.* Vol. 1. Canadian Wildlife Service Special Publication, no. CW66-96/1989-1E.

Nisbet, Ian C. T., Douglas B. McNair, William Post, and Timothy C. Williams. 1995. "Transoceanic Migration of the Blackpoll Warbler: Summary of Scientific Evidence and Response to Criticisms by Murray." *Journal of Field Ornithology* 66:612–622.

O'Brien, Michael, Richard Crossley, and Kevin Karlson. 2006. *The Shorebird Guide*. Boston: Houghton Mifflin Company.

Peterson, Roger Tory. 1959. *A Field Guide to the Birds*. Boston: Houghton Mifflin and Company.

Piersma, Theunis, Robert E. Gill Jr., Daniel R. Ruthrauff, Christopher G. Guglielmo, Jesse R. Conklin, and Colleen M. Handel. 2022. "The Pacific as the World's Greatest Theater of Bird Migration: Extreme Flights Spark Questions about Physiological Capabilities, Behavior, and the Evolution of Migratory Pathways." *Ornithology* 139:1–29.

Pitelka, F. A. 1950. "Geographic Variation and the Species Problem in the Shore-Bird Genus *Limnodromus*." *University of California Publications in Zoology* 50:1–108.

Ramirez, Vainey, and Claudio Delgado. n.d. "The Epic Voyage of the Hudsonian Godwit." Fundacion Conservacion Marina and WHSRN Executive Office. https://storymaps.arcgis.com/stories/1ebf7c221c654e6d8a06355501f40cb9.

Rantanen, Mika, Alexey Karpechko, Antti Lipponen, Kalle Nordling, Otto Hyvarinen, Kimmo Ruosteenoja, Timo Vihma, and Ari Laaksonen. 2022. "The Arctic Has Warmed Nearly Four Times Faster than the Globe Since 1979." *Communications Earth and Environment* 3:168.

Readfearn, Graham. 2022. "Bar-Tailed Godwit Sets World Record with 13,560 km Continuous Flight from Alaska to Southern Australia." *The Guardian*, October 22, 2022.

Robbins, Chandler S., Bertel Bruun, and Herbert S. Zim. 1966. *A Guide to Field Identification: Birds of North America*. New York: Golden Press.

Rowan, W. 1932. "The Status of the Dowitchers with a Description of a New Subspecies from Alberta and Manitoba." *Auk* 49:14–35.

Senner, Nathan R., Wesley M. Hochachka, James W. Fox, and Vsevolod Afanasyev. 2014. "An Exception to the Rule: Carry-Over Effects Do Not Accumulate in a Long-Distance Migratory Bird." *PLoS One* 9(2): e86588.

Senner, Nathan R., Maria Stager, and Brett K. Sandercock. 2016. "Ecological Mismatches Are Moderated by Local Conditions for Two Populations of a Long-Distance Migratory Bird." *Oikos* 126:61–72.

Senner, Nathan R., Maria Stager, Mo A. Verhoeven, Zachary A. Cheviron, Theunis Piersma, and Willem Bouten. 2018. "High-Altitude Shorebird Migration in the Absence of Topographic Barriers: Avoiding High Temperatures and Searching for Profitable Winds." *Proceedings of the Royal Society B* 285:20180569.

Shepherd, P. C. F., and J. S. Boates. 1999. "Effects of a Commercial Baitworm Harvest on Semipalmated Sandpipers and Their Prey in the Bay of Fundy Hemispheric Shorebird Reserve." *Conservation Biology* 13 (2): 347–356.

Sibley, David Allen. 2014. *The Sibley Guide to the Birds*. 2nd ed. New York: Alfred A. Knopf.

Smith, Paul A., Adam C. Smith, Brad Andres, Charles M. Francis, Brian Harrington, Christian Friis, R. I. Guy Morrison, Julie Paquet, Brad Winn, and Stephen Brown. 2023. "Accelerating Declines of North America's Shorebirds Signal the Need for Urgent Conservation Action." *Ornithological Applications* 125:1–14.

Stout, Gardner D. 1967. *The Shorebirds of North America*. New York: Viking.

Weidensaul, Scott. 2021. *A World on the Wing: The Global Odyssey of Migratory Birds*. New York: W. W. Norton.

Wiltschko, Roswitha, and Wolfgang Wiltschko. 2006. "Magnetoreception." *BioEssays* 28(2): 157–168.

Woodley, Keith. 2009. *Godwits: Long-Haul Champions*. Auckland: Penguin Group. Kindle.

Wynn, Joe, Oliver Padget, Henrik Mouritsen, Joe Morford, Paris Jaggers, and Tim Guilford. 2022. "Magnetic Stop Signs Signal a European Songbird's Arrival at the Breeding Site After Migration." *Science* 375:446–449.

Zust, Z., A. Mukhin, P. D. Taylor, and H. Schmaljohann. 2023. "Premigratory Flights in Migrant Songbirds: The Ecological and Evolutionary Importance of Understudied Exploratory Movements." *Movement Ecology* 11 (78): 1–15.

Index

(Page numbers in *italics* refer to illustrations; those in **bold** refer to species text boxes)

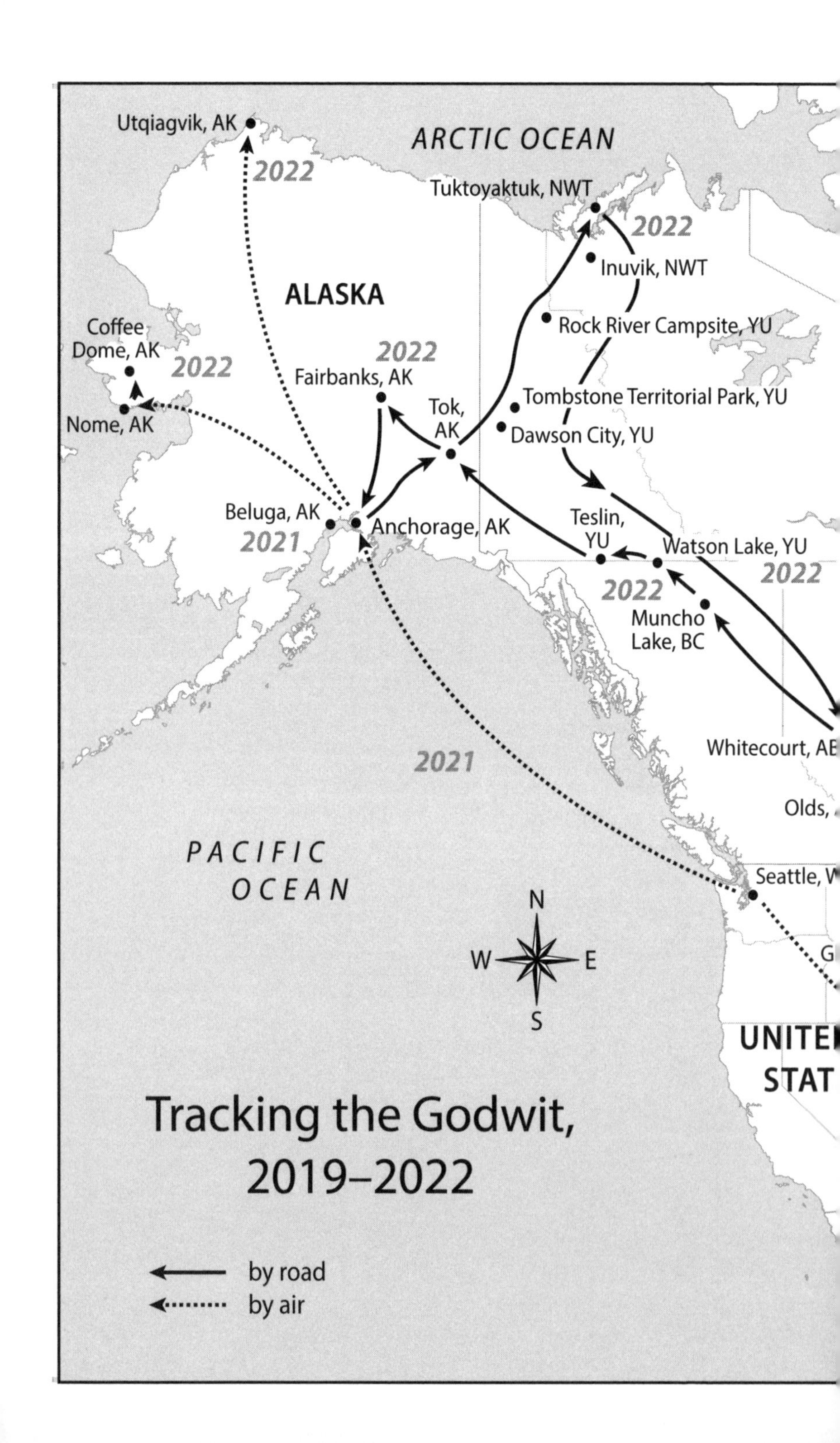
Tracking the Godwit,
2019–2022
by road
by air
ARCTIC OCEAN
PACIFIC
OCEAN
ALASKA
Utqiagvik, AK
Tuktoyaktuk, NWT
Inuvik, NWT
Rock River Campsite, YU
Coffee
Dome, AK
Nome, AK
Fairbanks, AK
Tok,
AK
Tombstone Territorial Park, YU
Dawson City, YU
Beluga, AK
Anchorage, AK
Teslin,
YU
Watson Lake, YU
Muncho
Lake, BC
Whitecourt,
Olds,
Seattle,
2022
2021
N
W
E
S